Actions of finite abelian groups

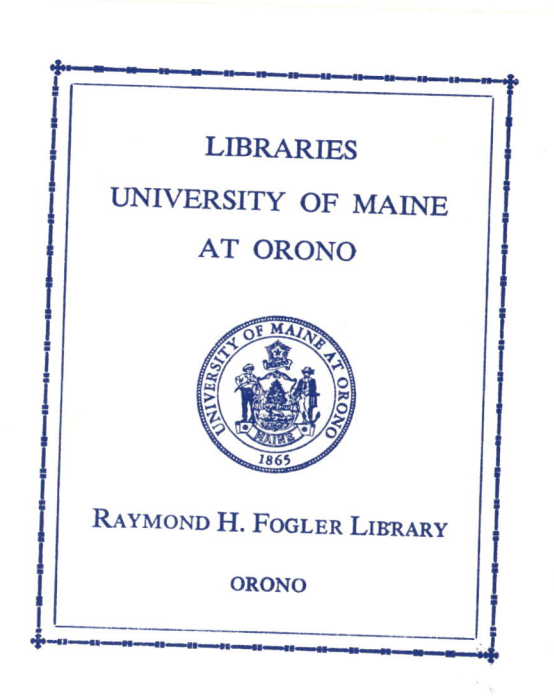

C Kosniowski
University of Newcastle upon Tyne

Actions of finite abelian groups

Pitman
LONDON · SAN FRANCISCO · MELBOURNE

PITMAN PUBLISHING LIMITED
39 Parker Street, London WC2B 5PB

FEARON-PITMAN PUBLISHERS INC.
6 Davis Drive, Belmont, California 94002, USA

Associated Companies
Copp Clark Ltd, Toronto
Pitman Publishing New Zealand Ltd, Wellington
Pitman Publishing Pty Ltd, Melbourne

First Published 1978

AMS Subject Classifications: (main) 57E-XX, 57D85
                              (subsidiary) 55B-XX, 55F35

© C Kosniowski 1978

All rights reserved. No part of this publication may be reproduced, stored in a retrieval system, or transmitted in any form or by any means, electronic, mechanical, photocopying, recording and/or otherwise without the prior written permission of the publishers. The paperback edition of this book may not be lent, resold, hired out or otherwise disposed of by way of trade in any form of binding or cover other than that in which it is published, without the prior consent of the publishers.

Reproduced and printed by photolithography
in Great Britain at Biddles of Guildford

ISBN 0 273 08405 4

Preface

This book is a contribution to the classification of smooth actions of groups on differentiable manifolds. The prerequistes have been kept to a minimum. In fact this book should be accessible to anyone who has a reasonably good background in

(i)   Algebraic Topology as in [Dold, 1], [Hilton and Wylie], [Spanier] etc.;

(ii) Cobordism theory as in [Stong, 1];

(iii) Morse theory as in [Hirsch] or [Milnor, 1]; and

(iv) Equivariant Topology as in [Bredon] and [Conner and Floyd, 1].

Results that are needed in this book but which are only available in research papers are usually given with new proofs. (This also includes results from books that rely heavily upon material in research papers.) Thus the hope is that the reader will only have to refer to a few books for background material and not to a string of journals. References to the original papers have been given in order to put the material into perspective - this has usually been left for the sections entitled "Historical Note".

Some exercises have been incorporated into the book and the reader will find that these exercises often provide alternative methods of proof of various results.

Very little of the material in this book can be found elsewhere. Much of it results from the author's research (hitherto unpublished) over the last year or so. A number of mathematicians have helped directly or indirectly with this book. I would foremost like to thank R.E. Stong for his interest and help in my work and also for providing me with the initial impetus. I would like to thank J.S. Rose for his help with many of the group theoretic problems that I encountered. Indeed many of the group theoretic results which appear here were shown to me by him. R. Lashof, W.D. Neumann, E. Ossa and R.E. Schultz have helped by their willingness to discuss some of the topological problems. I thank them for this. R.E. Schultz has also helped by allowing me to use some of his notes on G invariant Morse functions. I would also like to thank R.P. Beem for pointing out an error in an earlier draft of section 4.3.

I would like to thank the University of Newcastle upon Tyne and the University of Virginia - the former in part for allowing me to visit the latter, the latter for inviting me to visit.

My thanks to Marie Brown for her excellent typing, also to her husband, Charlie, for the drawings.

Finally I would like to thank my wife, Ann, and daughters, Kora and Inga, for being.

Contents

Introduction 1

Chapter 1: *Bordism Groups*

    1. Equivariant bordism groups 5

    2. Slice types 7

    3. Bordism with families of G slice types 9

    4. Reinterpretation of certain bordism groups 14

    5. Reinterpretation of the bundle bordism groups 17

    6. Irreducible G modules 19

    7. Splitting homomorphisms 24

    8. Surgery 31

    9. Historical note 45

Chapter 2: *Cutting and Pasting Groups*

    1. The SK groups 47

    2. Equivariant SK groups 48

    3. Bordism and SK equivalence 52

    4. Fibrations and SK equivalence 56

    5. Calculation of SK_* 60

    6. Splitting homomorphisms 62

    7. Bordism and splittings 66

    8. Historical note 69

Chapter 3: *Classifying Spaces*

    1. Bordism and cohomology   70

    2. Simplification of $\Gamma(H;U)$ 71

    3. Cohomology of $B\Gamma$ 80

    4. Generators of $N_*(B\Gamma)$ 98

    5. Calculation of $SK_*(B\Gamma)$ 114

    6. Historical note   114

Chapter 4: *Generators of the Equivariant Bordism Groups*

    1. Generators for G bordism, G odd order   116

    2. Generators for $\mathbb{Z}/2$ bordism   117

    3. Generators for $\mathbb{Z}/4$ bordism   120

    4. Generators for $\mathbb{Z}/2^n$ bordism   132

    5. $(\mathbb{Z}/2)^k$ bordism - algebraic results   145

    6. Generators of $(\mathbb{Z}/2)^k$ bordism   173

    7. Generators for G bordism, G general   180

    8. Generators of S^1 bordism   187

    9. Historical note   194

Chapter 5: *Generators of the Equivariant Cutting and Pasting Groups*

1. A general remark. 196
2. Generators of SK_*^G; G of odd order 197
3. Generators of $SK_*^{\mathbb{Z}/2}$ 200
4. Generators of $SK_*^{\mathbb{Z}/4}$ 207
5. Generators of SK_*^G; $G = \mathbb{Z}/2^r$ 215
6. Generators for SK_*^G; $G = (\mathbb{Z}/2)^k$ 217
7. Generators for SK_*^G; G general 220

References 222

Index 227

Introduction

An important area of topology is to understand, indeed classify, the actions of groups on differential manifolds. P.E. Conner and E.E. Floyd have demonstrated in their book [Conner and Floyd, 1] the effectiveness of bordism techniques in such an analysis. Another possible approach to the study of group actions is through the techniques of equivariant cuttings and pastings of manifolds. This leads to the so called SK groups (Sneiden and Kleben). Both techniques are related. Ways of producing new G manifolds are often through cutting and pasting techniques. Conversely results about bordism groups lead to results concerning the SK groups.

One of the purposes of this book is to study smooth actions of finite abelian groups on differential manifolds through the techniques of bordism and the techniques of cutting and pasting. Indeed a suitable "target" is: for a given group G describe a set of G manifolds which generate (or form a basis) of the G bordism ring N_*^G and describe another set that generate the cutting and pasting ring SK_*^G. If G is the trivial group then it is known that

$$N_* \cong (\mathbb{Z}/2)[x_2, x_4, x_5, \ldots]$$

the polynomial ring over $\mathbb{Z}/2$ on classes x_n of dimension n for each positive integer not of the form $2^m - 1$, (see section 1.1). Also,

$$SK_* \cong \mathbb{Z}[y_2]$$

the integral polynomial ring on a two dimensional class which may be taken to be the real projective plane RP^2, (see section 2.5). We wish to extend such results to other groups G.

Throughout this book we shall only be concerned with unoriented manifolds and with smooth actions of finite abelian groups. The material in the first two chapters can be very easily generalised to the case of a compact Lie group - often this just requires some additional conjugacy conditions and where appropriate this is mentioned.

We shall assume (usually) that all manifolds, vector bundles etc., are smooth, i.e., C^∞ differentiable. The word smooth will therefore be understood and usually omitted.

Acknowledgements have often been omitted from the main text and have been collected in a set of "historical notes" to be found at the end of every chapter except the last. These historical notes are not meant to be comprehensive, they are merely there to put the material into perspective.

CONVENTIONS. We shall often have need of k-tuples of integers such as

$$J = (j_1, j_2, \ldots, j_k)$$

which is sometimes more convenient typographically in the form

$$J = (j(1), j(2), \ldots, j(k)).$$

We shall freely use either of these alternatives and make no distinction between them. For example if $J = (j_1, j_2, \ldots, j_k)$ then x_J is given by

$$x_{j(1)} x_{j(2)} \cdots x_{j(k)} .$$

The groups G are always finite abelian. It is sometimes easier to use multiplicative notation and sometimes additive notation. We shall allow ourselves to use whichever notation is more convenient in a given situation.

NOTATION.

G: finite abelian group.

G *manifold*: a smooth manifold M together with a smooth action of G on M. By a smooth *action* of G on M we mean a group homomorphism $\Phi: g \to \phi_g$ from G to the group of diffeomorphisms of M. For each $m \in M$ and $g \in G$ we denote $\phi_g(m)$ simply by gm. Equivalently, a smooth *action* of G on M is a map (i.e. continuous function) $\phi: G \times M \to M$ with $\phi(g,m)$ abbreviated to gm such that (i) the map from M to M given by $m \to gm$ is a diffeomorphism for each $g \in G$, (ii) if $g, g' \in G$ then $g(g'm) = (gg')m$ for all $m \in M$, and (iii) if 1 denotes the identity element of G then $1m = m$ for all $m \in M$.

G_x = *isotropy subgroup* at $x \in M = \{g \in G ; gx = x\}$

G *module*: a finite dimensional real vector space together with a linear action of G on it.

$G(x)$ = *orbit* of $x \in M = \{gx ; g \in G\}$.

If M is an H manifold where H is a subgroup of G then the *balanced product* $G \times_H M$ is the cartesian product $G \times M$ factored by the equivalence relation $(g,x) \sim (gh, h^{-1}x)$ for

$h \in H$. The action of G on $G \times_H M$ is by left multiplication on G, i.e., $g'(g,x) = (g'g,x)$.

I: unit interval, identity matrix or a tuple of integers, the context usually being clear in any given situation.

O_k: group of real orthogonal $k \times k$ matrices.

U_l: group of complex unitary $l \times l$ matrices.

$\mathbb{Z}/k$: cyclic group of order k.

1 Bordism groups

1.1 EQUIVARIANT BORDISM GROUPS

One of the main objects of our study is N_*^G, the equivariant (unoriented) bordism ring of G manifolds. To define this ring we need the concept of G *bordism*. Two n dimensional G manifolds M_1, M_2 are said to be G *bordant* if there is an $(n+1)$ dimensional G manifold N whose boundary as a G manifold is the disjoint union M_1+M_2 of M_1 and M_2.

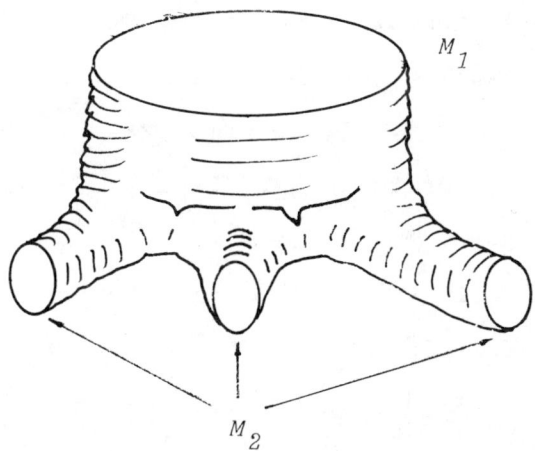

If M_1 and M_2 are G bordant then we usually write $M_1 \underset{G}{\sim} M_2$, or if no confusion can arise then we write simple $M_1 \sim M_2$. That "G bordism" is an equivalence relation on the set of closed n dimensional G manifolds is quite standard. We refer the reader to [Conner and Floyd, 1] and [Stong, 1]. The set of equivalences classes of n dimensional G manifolds is denoted by N_n^G. This becomes an abelian group if we define addition via disjoint

5

union of G manifolds. Defining $N_*^G = \oplus_{n \geq 0} N_n^G$ and using cartesian product of G manifolds as multiplication gives N_*^G a ring structure. If the group G is trivial then N_*^G becomes N_*, the unoriented bordism ring. In general N_*^G is an N_* module where multiplication is defined by cartesian product of manifolds. We shall be primarily interested in N_*^G as an N_* module.

Recall that N_* is the polynomial ring over $\mathbb{Z}/2$ on classes x_i of dimension i for each integer i not of the form 2^t-1. The even dimensional generators may be taken to be the classes of real projective spaces. Constructions for the odd dimensional generators were first given in [Dold, 2]. Another construction has been given in [Milnor, 2] - see also [Stong, 1]. A Dold manifold $P(m,n)$ is the quotient manifold of the free involution $(x,z) \to (-x,\bar{z})$ on $S^m \times CP^n$. If we write $2k = 2^r(2s+1)$, then the odd dimensional generator x_{2k-1} ($2k-1 \neq 2^t-1$) may be the class of $P(2^r-1, 2^r s)$.

The (Milnor) manifold $H_{m,n}$ ($m \leq n$) is the non-singular hypersurface of degree $(1,1)$ in $RP^m \times RP^n$. If we use homogeneous coordinates $[x_0, x_1, \ldots, x_m]$, $[y_0, y_1, \ldots, y_n]$ then $H_{m,n}$ is the set of points in $RP^m \times RP^n$ satisfying

$$x_0 y_0 + x_1 y_1 + \ldots + x_m y_m = 0.$$

If we write $2k = 2^r(2s+1)$ and let $m = 2^r$, $n = 2^{r+1}s$ then the odd dimensional generator x_{2k-1} ($2k-1 \neq 2^t-1$) may be the class of $H_{m,n}$.

This is perhaps a convenient point at which to remind the reader of the singular bordism groups. If X is a space then a singular n dimensional manifold in X is a pair (M,f) consisting of a compact n dimensional manifold M and a map $f:M \to X$. Strictly speaking a *singular n manifold* in X is an equivalence class (M,f) where (M_1,f_1) is equivalent to (M_2,f_2) if there is a diffeomorphism $M_1 \to M_2$ making the diagram

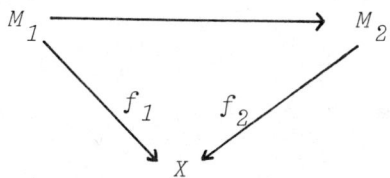

commute. Two such pairs (M,f), (M',f') are said to be *bordant* if there is an $(n+1)$ dimensional manifold N and a map $F:N \to X$ such that

(i) $\quad \partial N = M + N'$.

(ii) $\quad F|M = f, \quad F|M' = f'$.

The set of bordism classes of singular n manifolds in X forms an abelian group $N_n(X)$. Setting

$$N_*(X) = \bigoplus_{n \geq 0} N_n(X)$$

gives us an N_* module. See [Conner and Floyd, 1] for further details. We shall not need the singular G bordism groups, nor shall we have need of singular bordism of pairs of spaces.

1.2 SLICE TYPES

Let M be a G manifold and let $x \in M$. The *Slice Theorem* (see for instance [Bredon] or [Palais, 1,2]) tells us that there exists a G_x module $\bar{V}_x$ such that a suitable neighbourhood of the orbit

$G(x)$ of x in M is equivariantly diffeomorphic to $G \times_{G_x} \overline{V}_x$. The zero section G/G_x is mapped diffeomorphically onto the orbit $G(x)$ by $gG_x \to gx$. The subgroup G_x is in general only defined up to conjugacy and V_x is defined up to G_x equivalence of G_x modules. However, since G is abelian G_x and $\overline{V}_x$ are unique. We shall say that $\overline{V}_x$ is a G_x-$slice$ at x. The dimension of $\overline{V}_x$ as a real vector space is the same as the dimension of the G manifold M (assuming, as we are, that G is finite). The pair $[G_x; \overline{V}_x]$ is usually called the slice type at the point $x \in M$. For our purpose it will be convenient to look at the non-trivial part of $\overline{V}_x$. Thus $\overline{V}_x = V'_x \oplus V_x$ where G_x acts trivially on V'_x (in fact $V'_x = (\overline{V}_x)^{G_x}$) and no non-zero vector in V_x is fixed by all of G_x. We shall say that the pair $[G_x; V_x]$ is the $slice\ type$ of the point $x \in M$. Note that V_x determines $\overline{V}_x$ because $\overline{V}_x = R^{n-k} \oplus V_x$, where $n = \dim M$, $k = \dim V_x$ and G_x acts trivially on R^{n-k}. We say that $[H; U]$ is a G-$slice\ type$ if there is a G manifold M with $x \in M$ such that $G_x = H$ and $V_x = U$.

A $family$ of G-$slice\ types$ F is a collection of G-slice types $\{[H; U]; -\}$ satisfying the condition that if $[H;U] \in F$ and $x \in G \times_H U$ then the slice type $[G_x; U_x]$ at x belongs to F. Suppose that $F' \subseteq F$ are families of G-slice types. We say that the G manifold M is of $type\,(F, F')$ if for each $x \in M$ the slice type $[G_x; V_x] \in F$ and if for each $x \in \partial M$ the slice type $[G_x; V_x] \in F'$. If $F' = \emptyset$ then we talk of a G manifold of type F, which by the definition has to be a closed G manifold.

1.3 BORDISM WITH FAMILIES OF G-SLICE TYPES

By considering G manifolds of a certain "type" we obtain bordism groups which provide an inductive method of calculating N_*^G.

Suppose that $F' \subseteq F$ are families of G-slice types. Suppose that M_1 and M_2 are n dimensional G manifolds of type (F, F'). We say that M_1 and M_2 are (F, F') bordant if there exists an $(n+1)$ dimensional G manifold N of type (F, F) such that the disjoint union of M_1 and M_2 satisfies

$$M_1 + M_2 \subseteq \partial N \quad \text{(inclusion of } G \text{ submanifolds)}$$

and such that for each $x \in \partial N - (M_1 + M_2)$ the slice type $[G_x; V_x]$ at x belongs to the family F'.

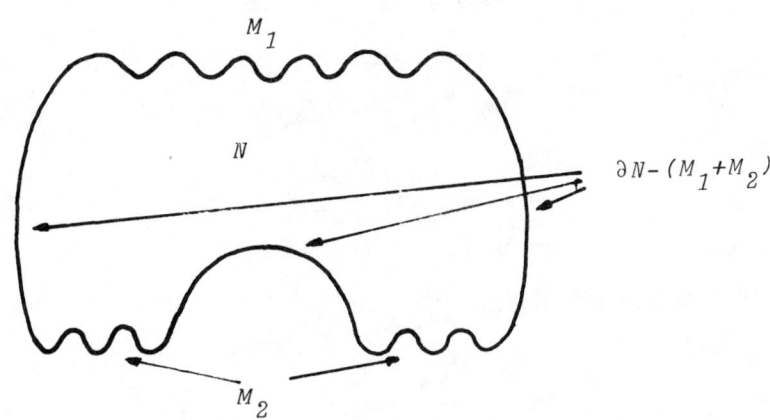

The relation "(F, F') bordant" is an equivalence relation on the set of closed n dimensional G manifolds of type (F, F'). The proof is quite analogous to that for other bordism theories as in [Conner and Floyd, 1] and [Stong, 1; Chapter 1]. The set of equivalence classes forms an abelian group under disjoint union.

It is denoted by $N_n^G[F,F']$. We set

$$N_*^G[F,F'] = \bigoplus_{n \geq 0} N_n^G[F,F']$$

which becomes an N_* module with multiplication defined by cartesian product of manifolds. If $F' = \emptyset$, then we write the resulting group as $N_*^G[F]$. If All denotes the family of all possible G-slice types then

$$N_*^G[All] = N_*^G \quad .$$

1.3.1 REMARK. A family F of subgroups of G is a collection of subgroups of G satisfying the condition that if $H \in F$ and $K \subseteq H$ then $K \in F$ (plus a conjugacy condition if G is non-abelian). If F is a family of subgroups of G then we denote by $All(F)$ the set of G slice types of the form $[H;U]$ with $H \in F$. If $F' \subseteq F$ is a pair of families of subgroups of G then we set

$$N_*^G(F,F') = N_*^G[All(F), All(F')] \quad .$$

The resulting groups agree with those defined in [Conner and Floyd, 2] and elsewhere via the notions of families of subgroups. (A G manifold M is of type $(All(F), All(F'))$ if and only if it satisfies the condition that if $x \in M$ then the isotropy subgroup $G_x \in F$ and if $x \in \partial M$ then $G_x \in F'$.)

There is an exact sequence of bordism groups relating triples $F'' \subseteq F' \subseteq F$ of families of G-slice types. Define N_* module homomorphisms

$$i: N_*^G[F',F''] \to N_*^G[F,F'']$$

$$j: N_*^G[F,F''] \to N_*^G[F,F']$$

by forgetting restrictions on slice types, (certainly a G manifold of type (F',F'') is also one of (F,F''), etc.). By assigning to a G manifold its boundary we get an N_* module homomorphism

$$\partial : N_*^G [F,F'] \to N_{*-1}^G [F',F'']$$

of degree -1.

1.3.2 THEOREM. *There is a long exact sequence*

$$\cdots \xrightarrow{\partial} N_n^G [F',F''] \xrightarrow{i} N_n^G [F,F''] \xrightarrow{j} N_n^G [F,F'] \xrightarrow{\partial} N_{n-1}^G [F',F''] \xrightarrow{i} \cdots$$

for any triple $F'' \subseteq F' \subseteq F$ *of families of G-slice types.*

Before proving this we shall prove a lemma which will be useful now and later on.

1.3.3 LEMMA. *Let M be an n dimensional G manifold of type* (F,F'). *Let W be a compact n dimensional G submanifold of M. Suppose that the slice type* $[G_x; V_x] \in F'$ *for all* $x \in M - W$. *Then M and W are* (F,F') *bordant.*

PROOF. Consider $M \times I$ with trivial action of G on I. The manifold $M \times I$ is of type (F,F') and

$$M + W = M \times \{0\} + W \times \{1\} \subseteq \partial(M \times I).$$

Finally, if $x \in (\partial(M \times I) - (M+W))$ then $[G_x; V_x] \in F'$.

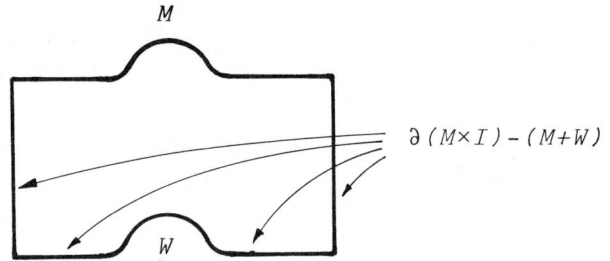

Thus $M \times I$ provides an (F,F') bordism between M and W. (An equivariant straightening of angles has to be made, see [Conner and Floyd, 1, pp. 9, 56].)

PROOF OF THEOREM. By the above lemma we see that $ji = 0$ (put $W = \emptyset$), so $\text{im}(i) \subseteq \ker(j)$. Suppose $M \in \ker(j)$, i.e., $j(M) = 0$. Thus there is an (F,F') bordism X between M and $\emptyset$. By definition X is of type (F,F) and $\partial X = M \cup Y$ where Y is of type (F',F''), since $\partial M = \partial Y$.

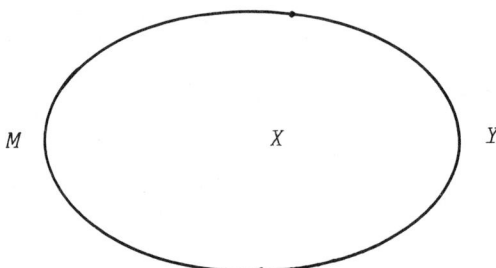

Make Y smaller by using an equivariant tubular neighbourhood N of its boundary so that $Y = N \cup Y'$ and $\partial X = M \cup N \cup Y'$.

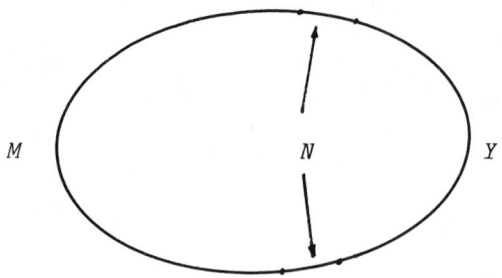

We see that X provides an (F,F'') bordism between M and Y', and Y' is an element in the image of i.

By use of the above lemma again we have $\partial j = 0$, i.e., $\text{im}(j) \subseteq \ker(\partial)$. Suppose $M \in \ker(\partial)$, i.e., $\partial M = 0$. Therefore, there is a G manifold X of type (F',F') with $\partial X = \partial M \cup Y$ and Y is of type $(F'',\emptyset)$, (thus Y is a closed G manifold).

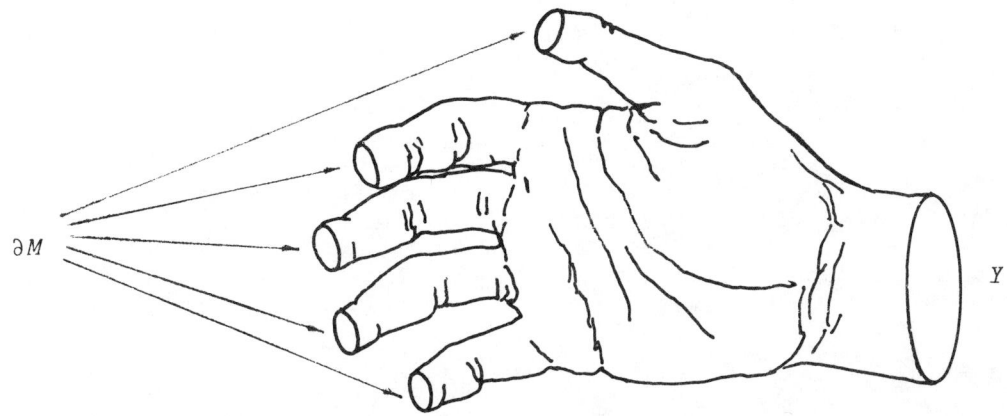

By gluing M and X along ∂M, we get a manifold, call it Z, which is in $\text{im}(j)$.

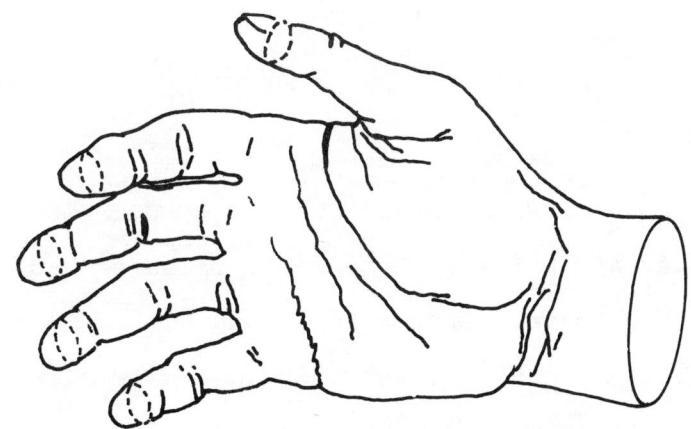

The manifold $Z \times I$ provides an (F,F') bordism between M and Z (using Lemma 1.3.3).

If M is of type (F,F') then M provides an (F,F') bordism between ∂M and $\emptyset$. Thus $i\partial = 0$, i.e., $\text{im}(\partial) \subseteq \ker(i)$. Suppose that $M \in \ker(i)$. Therefore there is a manifold X of type (F,F) with $M \subseteq \partial X$ and if $x \in \partial M$ then $[G_x; V_x] \in F''$.

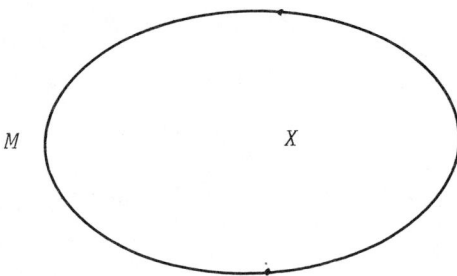

Consider $\partial X \times I$, Lemma 1.3.3 shows that this provides an (F',F'') bordism between M and ∂X.

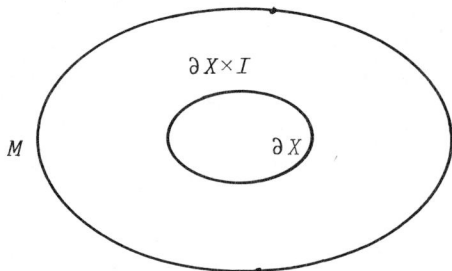

This completes the proof of exactness of the stated sequence.

1.4   REINTERPRETATION OF CERTAIN BORDISM GROUPS

For certain pairs $F' \subseteq F$ of families of G-slice types it is possible to reinterpret the group $N_*^G[F,F']$, essentially by using Lemma 1.3.3 and choosing W in some minimal way. The obvious choice for W (if it makes sense) is to take some small equivariant neighbourhood of the set N of all $x \in M$ having slice types $[G_x; V_x]$ in $F-F'$. In the case that N is a G submanifold then W can be taken as a small tubular neighbourhood of N -- this is the case in which we shall be interested. For our purpose it is sufficient to only consider the case in which $F-F'$ consists of a single slice type, say $[H;U]$. The reader can make the obvious generalizations.

If $F-F' = \{[H;U]\}$ and if M is of type (F,F') then the set N of all points $x \in M$ having slice type $[H;U]$ is a G submanifold of M and a small equivarint tubular neighbourhood of N is equivalent to a G vector bundle over N (the normal bundle of N in M). We are therefore led towards looking at bordism of G vector bundles. We say that a G vector bundle E over N is of type $[H;U]$ if the set of points in E having slice type $[H;U]$ is precisely N. Suppose that E_1, E_2 are vector bundles of type $[H;U]$ over compact closed manifolds N_1, N_2 respectively. We say that E_1 and E_2 are $[H;U]$ bordant if there exists a G vector bundle F of type $[H;U]$ over X such that

$$E_1 + E_2 \cong F|\partial X$$

where + denotes disjoint union and $F|\partial X$ as usual denotes the restriction of the vector bundle F to ∂X. The bundle bordism group $N_*^G[H;U]$ is the set of all $[H;U]$ bordism classes of G vector bundles of type $[H;U]$ over compact closed manifolds. The group structure is induced by disjoint union. $N_*^G[H;U]$ has a graded N_* module structure, where the grading is given by taking the dimension of the total space.

There is a natural N_* homomorphism

$$\nu_{[H;U]} : N_*^G[F,F'] \to N_*^G[H;U]$$

15

defined by assigning to M in $N_*^G[F,F']$ the normal bundle of N in M, where N is the set of points in M with slice type $[H;U]$. We leave the checking of details to the reader.

1.4.1 THEOREM. *If* $F-F' = \{[H;U]\}$ *then*
$$\nu_{[H;U]} : N_*^G[F,F'] \to N_*^G[H;U]$$
is an isomorphism.

PROOF. Define an N_*-homomorphism
$$N_*^G[H;U] \to N_*^G[F,F']$$
by assigning to a G vector bundle its unit disc bundle (with respect to some G invariant metric). Using Lemma 1.3.3 it is easy to see that this homomorphism is inverse to $\nu_{[H;U]}$.

1.4.2 COROLLARY. *If* $F-F' = \{[H;U]\}$ *then there is a long exact sequence*
$$\cdots \to N_n^G[F'] \xrightarrow{i} N_n^G[F] \xrightarrow{\nu} N_n^G[H;U] \xrightarrow{\partial} N_{n-1}^G[F'] \to \cdots$$
where $\nu(M)$ *is the normal bundle in M of the points in M having slice type* $[H;U]$ *and* $\partial(E)$ *denotes the unit sphere bundle of the G vector bundle E.*

1.4.3 EXERCISE. Let $F \subseteq F'$ be a pair of families of subgroups of G with $F-F' = \{H\}$. Prove that
$$N_*^G[All(F), All(F')] \cong \oplus N_*^G[H;U]$$
where the direct sum is taken over all slice types $[H;U]$ which are in $All(F)$ but not in $All(F')$. (See Remark 2.3.1 for the

relevant definitions.)

1.5 REINTERPRETATION OF THE BUNDLE BORDISM GROUPS

The N_* module $N_*^G[H;U]$ is isomorphic to a non-equivariant singular bordism group $N_{*-k}^G(B\Gamma(H;U))$ for some space $B\Gamma(H;U)$ where $k = \dim U$. The objective of this section will be to describe and prove this result.

The H module U determines in a natural way a map $r: H \to O_k$ where $k = \dim U$. Let $O(H;U)$ denote the centraliser of $r(H)$ in O_k.

1.5.1 DEFINITION. $\Gamma(H;U) = (G \times O(H;U))/\Delta$ where $\Delta = \{(h, r(h)); h \in H\}$.

Let us say that two G vector bundles E, E' of type $[H;U]$ are *isomorphic* if there is a vector bundle map $E \to E'$ which is a G equivariant diffeomorphism.

1.5.2 THEOREM. *The isomorphism classes of G vector bundles of type $[H;U]$ are in a one-to-one correspondence with the isomorphism classes of left principle $\Gamma(H;U)$ bundles.*

PROOF. Let P be the associated right principal O_k bundle of E, where $E = \dim U$. The group G acts on the left of P and the group O_k acts on the right. Let Q be the submanifold defined by

$$Q = \{p \in P;\ hp = pr(h) \text{ for all } h \in H\}$$

where $r: H \to O_k$ is the natural map determined by U.

The group $\Gamma(H;U)$ acts on Q (on the left) as follows

$$(g,\alpha)q = gq\alpha^{-1} \text{ for } g \in G, \alpha \in O(H;U), q \in Q$$

because

$$(gh, \alpha r(h))q = ghqr(h)^{-1}\alpha^{-1}$$
$$= gq\alpha^{-1}$$
$$= (g,\alpha)q$$

and

$$(g',\beta)((g,\alpha)q) = (g',\beta)(gq\alpha^{-1})$$
$$= g'gq\alpha^{-1}\beta^{-1}$$
$$= (g'g, \beta\alpha)q$$
$$= ((g',\beta)(g,\alpha))q.$$

Suppose that the element $(g,\alpha) \in \Gamma(H;U)$ fixes q, then $gq\alpha^{-1} = q$, i.e., $gq = q\alpha$. But if q is in the fibre over $x \in M$ then so is $q\alpha$ and so g fixes $x \in M$, hence $g \in H$. By definition of Q we then have $\alpha = r(g)$ and so $(g,\alpha) = (g,r(g)) = (1,1)$ since $(g,r(g)) \in \Delta$ as $g \in H$. Thus $\Gamma(H;U) = (G \times O(H;U))/\Delta$ acts freely on Q and Q is a left principal $\Gamma(H;U)$ bundle.

It is easy to see that isomorphic G vector bundles of type $(H;U)$ produce isomorphic principle $\Gamma(H;U)$ bundles.

Note that we can recover P (and hence E) from Q because

$$P = Q \times_{O(H;U)} O_k$$

where $O(H;U)$ acts on Q (on the right) simply as $(q)(\alpha) = q\alpha$ and, of course, $O(H;U)$ acts on O_k (on the left) via its inclusion $O(H;U) \subseteq O_k$. In general, if we start with Q and a left action of $\Gamma(H;U)$ on Q then there is a right action of $O(H;U)$ on Q given by

$$q \cdot \alpha = f(\alpha^{-1})q \qquad \alpha \in O(H;U), \ q \in Q$$

where

$$f : O(H;U) \to (G \times O(H;U))/\Delta = \Gamma(H;U)$$

is the homomorphism

$$\alpha \to (1,\alpha).$$

We see that

$$P = Q \times_{O(H;U)} O_k$$

is a right principal O_k bundle with a left action of G. The associated R^k vector bundle is a G vector bundle of type $[H;U]$.

It is easy to check the one-to-one correspondence now.

1.5.3 COROLLARY. $N^G_*[H;U] \cong N_{*-k}(B\Gamma(H;U))$ where $k = \dim U$. *If G is not finite then $k = \dim U + \dim G - \dim H$.*

1.6 IRREDUCIBLE G MODULES

In this section we shall describe the irreducible G modules and also calculate the number of inequivalent ones. This information will enable us to see what G slice types occur.

Let V be an irreducible G module, then either

$$\dim_R V = 1, \text{ and the underlying space is } R$$

or

$$\dim_R V = 2, \text{ and the underlying space is } C.$$

This follows essentially from Schur's lemma together with the fact that G is abelian (see for example [Adams]).

Suppose that H is a subgroup of G, such that the quotient G/H is cyclic. Then we shall define irreducible G modules

$$V(H,j) \quad 1 \leq j < (1/2)\phi(|G/H|) + 1$$

where ϕ is Euler's function, as follows. If $|G/H| = 1$ then $V(H,1)$ is R with trivial G action. If $|G/H| = 2$ then $V(H,1)$ is R with action of $g \in G$ given by multiplication by

$$\begin{cases} +1 & \text{if } g \in H \\ -1 & \text{if } g \in G-H. \end{cases}$$

If $|G/H| = q > 2$ then choose a decomposition of G/H into cosets $H \cup xH \cup \ldots \cup x^{q-1}H$ and let $m_1 < m_2 < \ldots < m_{\phi(q)}$ be a complete set of positive integers which are less than q and also prime to q (this incidentally defines Euler's ϕ function). Define $V(H,j)$ to be the complex numbers with action $g \in G$ given by multiplication by

$$\exp((2\pi i n m_j)/q) \quad \text{if } g \in x^n H \quad .$$

The H modules $V(H,j)$, $V(H,\phi(q) + 1-j)$ are isomorphic as H modules. The isomorphism is obtained by using conjugation on the underlying space

$$\begin{array}{ccc} C & \xrightarrow{\cong} & C \\ z & \longrightarrow & \bar{z} \end{array}$$

and noting that $m_{\phi(q)+1-j} = q - m_j$ so that multiplication by $\exp((2\pi i n m_j)/q)$ goes over to multiplication by $\exp((2\pi i n m_{\phi(q)+1-j})/q)$ after conjugation.

1.6.1 THEOREM. *Let G be a finite abelian group. The irreducible G modules are*

$$\{V(H;j); H \subseteq G, G/H \text{ cyclic}, 1 \leq j < (1/2)\phi(|G/H|) + 1\}$$

where ϕ is Euler's function.

To prove the theorem, suppose that V is a non-trivial irreducible G module. If V has underlying space R then the action of G induces a non-trivial homomorphism

$$\rho: G \to \{\pm 1\} = S^0 \subset R.$$

Therefore $\ker(\rho)$ is a subgroup of index 2 in G. The action of $g \in G$ on V is given by multiplication by $\rho(g)$, i.e., multiplication by

$$\begin{cases} 1 & \text{if } g \in H \\ -1 & \text{if } g \in G-H \end{cases}$$

Thus $V = V(H,1)$.

Now suppose that $\dim_R V = 2$. The action of G on V induces a non-trivial homomorphism

$$\rho: G \to S^1 \subset C.$$

If $H = \ker(\rho)$ then G/H is a cyclic group because

$$G/H \to S^1$$

is injective and finite subgroups of S^1 are cyclic. Furthermore

$$\rho(G/H) = \{\exp(2\pi i n/q); n = 0,1,\ldots,q-1\}$$

where $q = |G/H|$ and $g \in G$ acts on V by multiplication by $\rho(g)$ and of course

$$\rho(g) = \exp(2\pi i n m/q)$$

where m is prime to q and is determined by ρ while n depends on the coset of G/H to which g belongs (i.e., $g \in x^n H$ where $G/H = H \cup xH \cup x^2 H \cup \ldots \cup x^{q-1} H$). It follows that $V = V(H, m)$.

1.6.2 LEMMA. *The number of irreducible, inequivalent G modules is*

$$\frac{|G| + 2^d}{2}$$

where d is the minimum number of generators of the sylow 2-subgroup T of G, ($d = 0$ if $T = 1$).

PROOF. Let $V_0, V_1, \ldots, V_{r-1}$ be a complete set of irreducible inequivalent G modules, with V_0 being the trivial G module (i.e., R with trivial action of G). If $d_i = \dim V_i$ then

$$|G| = \sum_{i=0}^{r-1} d_i ,$$

see for example [Curtis and Reiner]. Let s be the number of G modules with R as underlying space, then $s-1$ is the number of subgroups of index 2 in G by the previous theorem. (This number $s-1$ is also the number of non-trivial homomorphisms $G \to S^0$.) If T is the sylow 2-subgroup of G then $s-1$ is the number of subgroups of index 2 in T. Let $\Phi(T)$ denote the intersection of all the subgroups of index 2 in T. We see that (i) $|T/\Phi(T)| = 2^d$ where d is the minimum number of generators of T, (ii) $s-1$ is the number of subgroups of index 2 in T and (iii) $T/\Phi(T)$ is an elementary abelian 2-group (i.e., of the form $\mathbb{Z}/2 \times \mathbb{Z}/2 \times \ldots \times \mathbb{Z}/2$) or order 2^d. Since the number of subgroups of index 2 in an elementary abelian 2-group of order 2^d is $2^d - 1$ we immediately have

$$s = 2^d.$$

Hence

$$|G| = s + 2(r-s)$$
$$= 2r - s$$
$$= 2r - 2^d$$

from which the required result follows.

1.6.3   REMARK.   The result can also be deduced by using a deep theorem of S.D. Berman and E. Witt (see [Curtis and Reiner; p. 306]).

Recall from section 1.5 that $O(H;U)$ is the centraliser of the image of $r(H)$ in O_k where $r: H \to O_k$ is the natural map determined by U and $k = \dim U$.

1.6.4   LEMMA.   *Suppose that U has a decomposition*

$$U = V_1^{k(1)} \times V_2^{k(2)} \times \ldots \times V_m^{k(m)} \times W_1^{l(1)} \times W_2^{l(2)} \times \ldots \times W_n^{l(n)}$$

where the V_i are one dimensional irreducible H modules and the W_i are two dimensional irreducible H modules. Suppose also that $V_i \neq V_j$ if $i \neq j$ and $W_i \neq W_j$ if $i \neq j$, then

$$O(H;U) \cong \prod_{i=1}^{m} O_{k(i)} \times \prod_{i=1}^{n} U_{l(i)} \quad .$$

The proof is quite straightforward.

The description of the map $r: H \to O(H;U) \cong \prod O_{k(i)} \times \prod U_{l(j)}$ is as follows:

(i) Given $O_{k(i)}$, there is a subgroup K_i of H of index 2 such that the composite

$$H \to O(H;U) \to O_{k(i)}$$

(also denoted by r) is given by

$$r(h) = \begin{cases} +I & \text{if } h \in K_i \\ -I & \text{if } h \in H-K_i \end{cases}$$

(ii) Given $U_{l(j)}$, there is a subgroup L_j of H of index $q \neq 2$ with H/L_j cyclic. Also there is a decomposition of H/L_j

$$L_j \cup xL_j \cup \ldots \cup x^{q-1}L_j$$

and a primitive q-th root of unity ξ_j such that the composition

$$H \to O(H;U) \to U_{l(j)}$$

(denoted by r) is given by

$$r(h) = (\xi_j)^k I \quad \text{if } h \in x^k L.$$

1.7 SPLITTING HOMOMORPHISMS

Let $F' \subseteq F$ be families of G slice types with $F-F' = \{[H;U]\}$. By placing further restrictions on F it is possible to define an N_* homomorphism

$$q : N_*^G[H;U] \to N_*^G[F]$$

which sometimes is inverse to v.

Suppose that E represents an element of $N_*^G[H;U]$. Then E is a G vector bundle over N say. Multiplying E by R gives another vector bundle $R \times E$ over N. We denote by $RP(R \times E)$ the real projective space bundle associated to the vector bundle $R \times E$. Since N is a manifold so is $RP(R \times E)$. In fact if E has fibre V then $RP(R \times E)$ is a fibre bundle over N with fibre $RP(R \times V)$. The action of G on E (trivial on R) induces an action of G on $RP(R \times E)$. Thus $RP(R \times E)$ is a G manifold. (See section 2.2 of [Atiyah].)

An alternative way of describing this manifold is to take $D(E)$, the unit disc bundle of E and place an equivalence relation on it which is non-trivial only on the boundary of $D(E)$ (the sphere bundle). On the sphere bundle the equivalence relation is $x \sim -x$ in each sphere of the sphere bundle. The resulting equivalence classes form a G manifold equivalent to $RP(R \times E)$.

If for all $x \in RP(R \times E)$ the slice type $[G_x; V_x]$ is contained in F then, of course, $RP(R \times E)$ is a G manifold of type F. In fact for this to be true it is sufficient that for all $x \in G \times_H RP(R \times U)$ the slice type $[G_x; V_x]$ belongs to F where $RP(R \times U)$ is the real projective space of $R \times U$ (an H vector bundle over a point). This is because the action of H on $R \times E$ is completely within each fibre of $R \times E$. Hence the action of H in $RP(R \times E)$ is within each fibre and so the slice types of $RP(R \times E)$ are the slice types of $G \times_H RP(R \times U)$.

The slice types of $G \times_H RP(R \times U)$ are of the following form

$$[K; W] , [K; NT(V \otimes (W \times R))]$$

where $[K; W]$ is a slice type of $G \times_H U$, the module $V \subseteq W$ is of real dimension 1 and NT denotes taking the non-trivial part of a module. That this is true can be quite easily seen from the definition of $RP(R \times U)$.

If K is a subgroup of H of index 2 then let V_K be the H module with underlying space R and action of $h \in H$ given by multiplication by

$$\begin{cases} 1 & \text{if } h \in K \\ -1 & \text{if } h \notin K . \end{cases}$$

1.7.1 THEOREM. *If there does not exist a subgroup K of H of index 2 such that $U = V_k \oplus W$ with $V_K \not\subseteq W$ and $V_K \otimes W = W$ then*

$$\nu_{[H;U]}(RP(R \times E)) = E$$

for any G vector bundle E of type $[H;U]$.

PROOF. We think of $G \times_H RP(R \times U)$ as

$$(G \times_H D(U))/\sim$$

where $x \sim -x$ if $x \in G \times_H S(U)$. Let $x \in G \times_H RP(R \times U)$. If $x \notin G \times_H S(U)$ then the slice type at x is a slice type of $G \times_H U$. If $x \in G \times_H S(U)$ then let $[G_x; U_x]$ be the slice type at x. The group G_x sends x to x or x to $-x$. In the first case $[G_x; U_x]$ must be a slice type of $G \times_H U$ since points in a neighbourhood of x are then invariant under G_x. In the second case, since G_x sends x to $-x$, it maps the line L joining x and $-x$ to itself. This line L determines a one dimensional H module V_K for some subgroup K of index 2 in G_x. Thus we have

$$U|G_x = V_x \oplus W$$

for some W, and

$$U_x = NT(V_K \otimes W) \oplus V_K$$

where *NT* denotes taking the non-trivial part.

By the conditions assumed on H we see that therefore the only points with slice type $[H;U]$ in $G \times_H RP(R \times U)$ are precisely those coming from the zero section of U. Hence also the only points in $RP(R \times E)$ with slice type $[H;U]$ are precisely those coming from the zero section of E. The result follows immediately.

1.7.2 REMARK. The conditions of Theorem 1.7.1 are satisfied if G is a group of odd order since there are then no subgroups of index 2. We shall return to this case shortly.

Consider the case in which $G \times_H RP(R \times U)$ is a G manifold of type F. Then $RP(R \times E)$ is a G manifold of type F for all $E \in N_*^G[H;U]$. Furthermore, if E and E' are $[H;U]$ bordant G vector bundles of type $[H;U]$ then there is a G vector bundle F over X of type $[H;U]$ such that

$$E + E' = F|\partial X$$

Since $RP(R \times F)$ is a G manifold of type (F,F) with boundary $RP(R \times E) + RP(R \times E')$ we see that there is a well defined function

$$N_*^G[H;U] \longrightarrow N_*^G[F]$$
$$E \longmapsto RP(R \times E) \; .$$

We leave it for the reader to check that this is in fact an N_* module homomorphism.

1.7.3 COROLLARY. *Let* $F' \subseteq F$ *be families of G slice types with* $F-F' = \{[H;U]\}$ *and suppose that the following two conditions hold:*

(i) $G \times_H RP(R \times U)$ is a G manifold of type F, and

(ii) there does not exist a subgroup K of H of index 2 such that $U = V_K \oplus W$ with $V_K \not\subseteq W$ and $V_K \otimes W = W$.

Then the sequence

$$0 \to N_*^G[F'] \xrightarrow{i} N_*^G[F] \xrightarrow{\nu} N_*^G[H;U] \to 0$$

is a short split exact sequence with splitting homomorphism $q: N_*^G[H;U] \to N_*^G[F]$ given by $q(E) = RP(R \times E)$ for $E \in N_*^G[H;U]$.

Thus we see that if the conditions of the corollary hold then

$$N_*^G[F] \cong N_*^G[F'] \oplus N_*^G[H;U] .$$

The conditions in the corollary do not always hold, however if G is of odd order then they do.

1.7.4 COROLLARY. *If G is a group of odd order then*

$$N_*^G \cong \bigoplus_{[H;U] \in St(G)} N_*^G[H;U]$$

where $St(G)$ denotes the set of G slice types. An isomorphism $\oplus N_*^G[H;U] \to N_*^G$ is given by $\oplus E_{[H;U]} \to \Sigma RP(R \times E_{[H;U]})$.

PROOF. Let $F' \subseteq F$ be a pair of families of G slice types with $F - F' = \{[H;U]\}$. If $E \in N_*^G[H;U]$ then $RP(R \times E)$ has the same slice types as $G \times_H RP(R \times U)$. But the slice types of $G \times_H RP(R \times U)$ are the same as those of $G \times_H U$ since U is a complex H module, (H must have odd order). Thus $RP(R \times E)$ is of type F. We deduce from Corollary 1.7.2 that

$$N_*^G[F] \cong N_*^G[F'] \oplus N_*^G[H;U] .$$

Next we show that there is a sequence of families of G slice types

$$F_0 \subseteq F_1 \subseteq F_2 \subseteq \ldots$$

such that $F_i - F_{i-1} = \{[H_i; U_i]\}$ for some G slice type $[H_i; U_i]$. In order to do this we shall order the set $St(G)$ of G slice types. We order as follows (the method is quite general):

(i) $[H; U] < [K; V]$ if dim U < dim V.

(ii) Suppose that dim U = dim V, then $[H; U] < [K; V]$ if $H < K$ in some ordering of the subgroups of G preserving inclusion of subgroups, i.e., if $H \subseteq K$ then $H \leq K$. Such orderings do exist.

(iii) Suppose that dim U = dim V and $H = K$, then $[H; U] < [H; V]$ if U precedes V in the ordering of H modules induced lexicographically from some ordering on the irreducible H modules. Indeed, we see from Theorem 1.6.1 that an ordering of the subgroups of G (and hence an ordering of the subgroups of H, $H \subseteq G$) leads to an ordering of the irreducible H modules. The irreducible H module $V(K_1, i)$ precedes $V(K_2, j)$ if $K_2 < K_1$ or $K_2 = K_1$ and $i < j$.

We call the G slice types so ordered

$$\rho_0, \rho_1, \rho_2, \ldots$$

so that if $i < j$ then $\rho_i < \rho_j$. Define F_j, for $j \geq 0$, by

$$F_j = \{\rho_0, \rho_1, \ldots, \rho_j\} .$$

It is clear that each of F_j is a family of G slice types and that in fact

$$N_n^G [F_N] \cong N_n^G$$

if N is sufficiently large compared to n. Furthermore, $F_j - F_{j-1}$ consists of a single G slice type; thus the corollary easily follows.

1.7.5. COROLLARY. *If G is a finite abelian group of odd order then*

$$N_*^G \cong \bigoplus_{[H;U] \in St(G)} N_{*-|U|}(B\Gamma(H;U))$$

where $|U| = dim\ U$.

In general, we order the G slice types in a way as described for the case of G of odd order. We denote the G slice types by

$$\rho_0, \rho_1, \rho_2, \ldots$$

so that if $i < j$ then $\rho_i < \rho_j$ in the ordering of G slice types. The sets

$$F_j = \{\rho_0, \rho_1, \ldots, \rho_j\}$$

are families of G slice types and $F_j - F_{j-1} = \{\rho_j\}$. In general we have

$$N_n^G[F_N] \cong N_n^G$$

if N is sufficiently large compared to n. In certain circumstances we have

$$N_*^G[F_j] \cong N_*^G[F_{j-1}] \oplus N_*^G[\rho_j]$$

for example if $\rho_j = [H;U]$, $F_j = F$ satisfy the two conditions of Corollary 1.7.3. We analyse the situation more fully in Chapter 4.

1.8 SURGERY

There is an alternative way of looking at bordism and this is via the concept of *surgery*. This will be used later on in our calculations of the *SK* groups.

Suppose that M is an n dimensional G manifold of type F and suppose that V_1, V_2 are H modules (for some subgroup $H \subseteq G$) with $\dim V_1 + \dim V_2 = n+1$. If G is not finite then the condition is that $\dim V_1 + \dim V_2 + \dim G - \dim H = n+1$. Suppose furthermore that there exists an equivariant regular embedding

$$\phi: G \times_H (S(V_1) \times D(V_2)) \to M \ .$$

We can form

$$W = M \times I \ \cup \ (G \times_H (D(V_1) \times D(V_2)))$$

by identifying $(g,x,y) \ \varepsilon \ G \times_H (S(V_1) \times D(V_2))$ with $(\phi(g,x,y),1)$ $\varepsilon \ M \times I$.

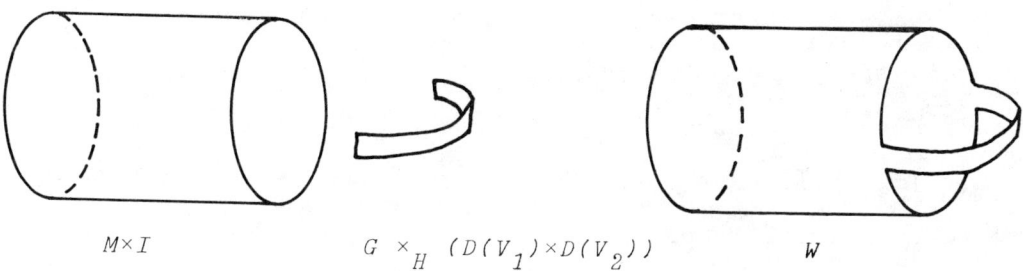

$M \times I$ $\qquad\qquad G \times_H (D(V_1) \times D(V_2)) \qquad\qquad W$

After smoothing, W is a G manifold with boundary $\partial W = M + M'$. If $G \times_H (D(V_1) \times D(V_2))$ is a G manifold of type (F,F) then M' is a G manifold of type F. We say that M' results from M by a *surgery of type* $[H, V_1, V_2; F]$. The *trace* of the surgery is by definition W. We can write M' in terms of M as

$$M' = (M - \phi(G \times_H (S(V_1) \times D(V_2)))) \cup_\phi G \times_H (D(V_1) \times S(V_2)).$$

We shall often omit the family F from our notation $[H_1, V_1, V_2; F]$.

As an example of a surgery observe that $G \times_H (S(V_1 \times R) \times S(V_2))$ results from $G \times_H S(V_1 \times V_2)$ by a surgery of type $[H, V_1, V_2]$ (the family required is that containing the slice types of $G \times_H D(V_1 \times V_2)$. This can be seen as follows

$$G \times_H (S(V_1 \times R) \times S(V_2)) = G \times_H (D(V_1) \times S(V_2)) \cup G \times_H (D(V_1) \times S(V_2))$$

$$= G \times_H (S(V_1 \times V_2) - (S(V_1) \times D(V_2)))$$

$$\cup G \times_H (D(V_1) \times S(V_2))$$

$$= (G \times_H S(V_1 \times V_2) - G \times_H (S(V_1) \times D(V_2)))$$

$$\cup G \times_H (D(V_1) \times S(V_2))$$

with the obvious identifications.

If M is a G manifold of type F and M' results from M by a surgery of type $[H, V_1, V_2; F]$ then M and M' are F bordant clearly. If we used a sequence of such surgeries then again the resulting manifolds would be F bordant. The next result states that the converse is also true.

1.8.1 THEOREM. *Let F be a family of G slice types. Two manifolds M, M' of type F are F bordant if and only if M' can be obtained from M by a sequence of surgeries of type* $[H_i, V_{i,1}, V_{i,2}; F]$ $i = 1, 2, \ldots, r.$

This result follows from the existence of invariant Morse functions. For completeness sake an elementary proof of this is given.

Denote the tangent space of a smooth G manifold M at a point p by TM_p. If $f: M \to N$ is a smooth G map between G manifolds with $f(p) = q$, then the induced linear map of tangent spaces will be denoted by $f_*: TM_p \to TN_q$.

Suppose that $f: M \to R$ is a smooth G invariant real valued function on M.

1.8.2 DEFINITION. A point $p \in M$ is called a *critical point* of f if the induced map
$$f_*: TM_p \to TR_{f(p)}$$
is zero.

If we choose a G invariant local coordinate system $(x_1, x_2, \ldots, x_n)$ in a G invariant neighbourhood of p then the condition in the definition is equivalent to
$$\frac{\partial f}{\partial x_1}(p) = \frac{\partial f}{\partial x_2}(p) = \ldots = \frac{\partial f}{\partial x_n}(p) = 0$$

It follows from the definition that each point in the orbit $G(p)$ of a critical point p is also a critical point. We can thus talk about *critical orbits*.

1.8.3 DEFINITION. The *Hessian* $H(f)_p$ of f at the critical point p is defined (with respect to some G invariant metric) to be the (symmetric) matrix

$$H(f)_p = \frac{\partial^2 f}{\partial x_i \partial x_j}(p) \quad .$$

The Hessian defines a linear transformation of TM_p. In the case that G is finite we say that the critical orbit $G(p)$ is non-degenerate if and only if $H(f)_p$ is non-singular, i.e., if

$$H(f)_p : TM_p \to TM_p$$

is invertible. In general we say that the critical orbit $G(p)$ is *non-degenerate* if

$$\operatorname{rank}(H(f)_p) = \dim(M) - \dim(G(p)).$$

This is because $H(f) = 0$ on $TG(p)$ since f_* is zero on $TG(p)$. We thus get a linear transformation of $(TM_p)/TG(p)_p$. An equivalent definition for the critical orbit $G(p)$ to be non-degenerate is that the linear transformation of $(TM_p)/TG(p)_p$ is invertible. The definitions for non-degeneracy are independent of the local coordinate system chosen.

We now state and prove the *Equivariant Morse Lemma*. Recall that a G_x slice V at x is a G_x module V such that an invariant neighbourhood of $G(x)$ is equivariantly diffeomorphic to $G \times_{G_x} V$. The module V is an orthogonal G_x space with respect to some G invariant metric.

1.8.4 PROPOSITION. *Suppose that M is compact and that $G(p) \subseteq M$ is a non-degenerate critical orbit for some smooth invariant function $f : M \to R$. Let $H = G_p$. If V is an H slice at p then there exists an H equivariant map $P : V \to V$ with $P^2 = P$ and $\langle Px, x' \rangle = \langle x, Px' \rangle$. Furthermore there exists a G diffeomorphism*

Θ *of an invariant neighbourhood of* p *to* $G \times_H V$ *such that*

$$f\Theta^{-1}(g,x) = f(p) + |P(x)|^2 - |x-P(x)|^2 .$$

The condition $\langle Px, x'\rangle = \langle x, Px'\rangle$ is equivalent to the condition that the matrix of P is symmetric.

PROOF. By the slice theorem the above reduces to looking at $f|V$, so the proof reduces to considering $G = H$, $M = V$, $p = 0$. We may as well assume that $f(0) = 0$ also.

Let $g_{i,j}(x) = g_{i,j}(x_1, \ldots, x_n)$ be given by

$$g_{i,j}(x) = \int_{I \times I} \frac{\partial^2 f}{\partial x_i \partial x_j} (stx) t \, ds \, dt$$

then

$$f(x) = \sum g_{i,j}(x) x_i x_j .$$

That this is true can be easily seen as follows:

$$\sum g_{i,j}(x) x_i x_j = \sum \int_{I \times I} \frac{\partial^2 f}{\partial x_i \partial x_j} (stx) x_i x_j t \, ds \, dt$$

$$= \sum_j \int_I t \int_0^1 (\sum_i \frac{\partial^2 f}{\partial x_i \partial x_j} (stx_1, \ldots, stx_n) x_i x_j ds) dt$$

$$= \sum_j \int_I \int_0^1 \frac{d}{ds} \frac{\partial f}{\partial x_j} (stx_1, \ldots, stx_n) ds \, x_j \, dt$$

$$= \sum_j \int_0^1 \frac{\partial f}{\partial x_j} (tx_1, \ldots, tx_n) x_j \, dt$$

$$= \int_0^1 \frac{d}{dt} f(tx_1, \ldots, tx_n) dt$$

$$= f(x).$$

We may therefore write $f(x)$ as

$$f(x) = \langle A(x)x, x\rangle$$

35

where $A(x)$ is the matrix $(g_{i,j}(x))$. Observe that $A(x)$ is symmetric and $g^{-1}A(gx)g = A(x)$. It is easy to check that

$$\frac{\partial^2}{\partial x_i \partial x_j} f(0) = g_{i,j}(0)$$

and hence by the non-degeneracy condition of f we see that $A(0)$ is a non-singular matrix.

We want a map B from a neighbourhood of 0 to the set of non-singular matrices which satisfies the condition $B(x)^{tr}A(0)B(x) = A(x)$, where tr denotes the transpose of a matrix. Consider

$$B(x) = (A(0)^{-1}A(x))^{\frac{1}{2}} = (I-(I-A(0)^{-1}A(x)))^{\frac{1}{2}}$$

where the latter is expanded by the binomial series for $|x|$ small. Thus $B(x)$ is some power series in $(I-A(0)^{-1}A(x))$. To see that this works let $C(x) = A(0)^{-1}A(x)$, then since $A(x)$ is symmetric we have

$$C(x)^{tr} = A(x)A(0)^{-1}$$
$$C(x)^{tr}A(0) = A(0)C(x)$$
$$B(x)^{tr}A(0) = A(0)B(x)$$

since $B(x)$ is a power series in $I-C(x)$. Hence

$$B(x)^{tr}A(0)B(x) = A(0)B(x)^2$$
$$= A(0)C(x)$$
$$= A(x) .$$

We now define an equivariant diffeomorphism $\overline{\Theta}$ by $\overline{\Theta}(x) = B(x)x$. To see that it is a diffeomorphism we observe that if $B(x) = (b_{i,j}(x))$ then

$$\left(\frac{\partial}{\partial x_1}, \frac{\partial}{\partial x_2}, \ldots, \frac{\partial}{\partial x_n}\right) \overline{\Theta}(x)\bigg|_{x=0}$$

$$= \left(\frac{\partial}{\partial x_i} \left(\sum_k b_{j,k}(x) x_k\right)\right)\bigg|_{x=0}$$

$$= (b_{j,i}(x))\bigg|_{x=0}$$

$$= B(0)^{tr}$$

$$= I.$$

We prove that $\overline{\Theta}$ is equivariant as follows: We have

$$g^{-1} A(gx) g = A(x)$$

or

$$A(gx) = g A(x) g^{-1}$$

and hence

$$C(gx) = g C(x) g^{-1}$$

since g commutes with $A(0)$ and hence with $A(0)^{-1}$. Using the power series expansion of $B(x)$ we obtain

$$B(gx) = g B(x) g^{-1}$$

and finally

$$\overline{\Theta}(gx) = B(gx) gx = g B(x) g^{-1} gx = g B(x) x = g \overline{\Theta}(x).$$

Hence $\overline{\Theta}$ is equivariant. Note that since $f(x) = \langle A(x)x, x \rangle$ we have

$$f(x) = \langle B(x)^{tr} A(0) B(x) x, x \rangle$$

$$= \langle A(0) B(x) x, B(x) x \rangle$$

$$= \langle A(0) \overline{\Theta}(x), \overline{\Theta}(x) \rangle.$$

Because $A(0)$ is a non-singular symmetric matrix we may write it, after an orthogonal change of basis, as

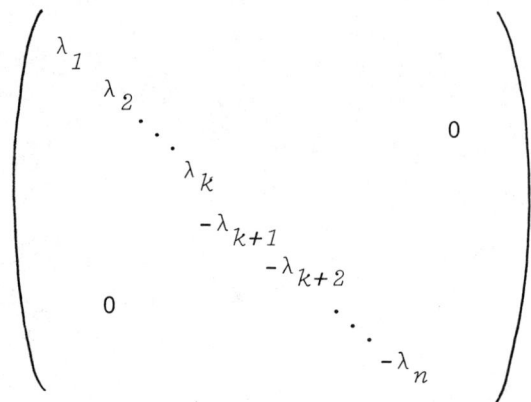

where each λ_i is a positive real number. Let A^+ be given by

$$\begin{pmatrix} \lambda_1 & & & 0 \\ & \lambda_2 & & \\ & & \ddots & \\ 0 & & & \lambda_k \end{pmatrix}$$

and A^- be given by

$$\begin{pmatrix} \lambda_{k+1} & & 0 \\ & \ddots & \\ 0 & & \lambda_n \end{pmatrix} .$$

Define Θ by

$$\Theta = (\sqrt{A^+} - \sqrt{A^-})\overline{\Theta}$$

where by $(\sqrt{A^+} - \sqrt{A^-})$ we mean the matrix

$$\begin{pmatrix} \sqrt{A^+} & 0 \\ 0 & -\sqrt{A^-} \end{pmatrix} .$$

Clearly Θ is an equivariant diffeomorphism. Let $P: V \to V^+$ be the projection of V onto V^+ the subspace of V determined by the eigenspaces corresponding to the eigenvalues $\lambda_1, \lambda_2, \ldots, \lambda_k$. Obviously $P^2 = P$ and $<Px, x'> = <x, Px'>$. The fact that P is equivariant follows from the fact that $gA(0) = A(0)g$. If V^- denotes the subspace of V corresponding to the negative eigenvalues $(-\lambda_{k+1}, \ldots, -\lambda_n)$ then $V = V^+ \oplus V^-$ and $I-P$ is a projection from V to V^-. It is clear that $PA(0)(x) = A^+P(x)$ and that $(I-P)A(0)(x) = A^-(I-P)(x)$.

To complete the proof of the proposition we calculate:
$$f(x) = <A(0)\overline{\Theta}(x), \overline{\Theta}(x)>$$
$$= <(P+I-P)A(0)\overline{\Theta}(x), (P+I-P)\overline{\Theta}(x)>$$
$$= <PA(0)\overline{\Theta}(x), P\overline{\Theta}(x)> + <(I-P)A(0)\overline{\Theta}(x), (I-P)\overline{\Theta}(x)>$$

(the other terms are zero since $V = V^+ \oplus V^-$)
$$= <A^+P\overline{\Theta}(x), P\overline{\Theta}(x)> - <A^-(I-P)\overline{\Theta}(x), (I-P)\overline{\Theta}(x)> \ .$$

Hence we have
$$f\overline{\Theta}^{-1}(x) = <A^+P(\sqrt{A^+} - \sqrt{A^-})^{-1}x, P(\sqrt{A^+} - \sqrt{A^-})^{-1}x>$$
$$- <A^-(I-P)(\sqrt{A^+} - \sqrt{A^-})^{-1}x, (I-P)(\sqrt{A^+} - \sqrt{A^-})^{-1}x> \ .$$

But
$$P(\sqrt{A^+} - \sqrt{A^-})^{-1}x = (\sqrt{A^+})^{-1}Px$$

and
$$(I-P)(\sqrt{A^+} - \sqrt{A^-})^{-1}x = (-\sqrt{A^-})^{-1}(I-P)x \ .$$

Thus
$$f\overline{\Theta}^{-1}(x) = <A^+(A^+)^{-\frac{1}{2}}Px, (A^+)^{-\frac{1}{2}}Px>$$
$$- <A^-(A^-)^{-\frac{1}{2}}(I-P)x, (A^-)^{-\frac{1}{2}}(I-P)x>$$
$$= <Px, x> + <(I-P)x, (I-P)x>$$

which completes the proof of the proposition.

1.8.5 COROLLARY. *The set of non-degenerate critical orbits in M is isolated (hence finite).*

Before we can prove Theorem 1.8.1 we need to show the existence of G invariant functions with non-degenerate critical orbits.

1.8.6 DEFINITION. A smooth G invariant function $f: M \to R$ is an *invariant Morse function* if it has no critical points on ∂M and all critical points lie on non-degenerate critical orbits. In addition it is standard to assume that f is constant on each component of ∂M. By Corollary 1.8.5 there are only finitely many critical orbits.

Let $ND_G(A, M)$ be the set of all invariant smooth functions $M \to R$ that are invariant Morse functions near A. If A is compact then $ND_G(A, M)$ is open in the set $C^\infty(M)^G$ of invariant smooth functions $M \to R$. (This is clear: either the first derivative is non-zero or the matrix of second partials has non-zero determinant, in either case the same holds for nearby functions.)

1.8.7 THEOREM. $ND_G(M, M)$ *is dense in* $C^\infty(M)^G$.

PROOF. We take the result as known for $G = 1$, see for example [Milnor, 1] or [Hirsch]. We prove the result by induction on the subgroups of G. So suppose that the result holds for every closed proper subgroup of G.

Notice that if S_x is a closed G_x slice at x then

$$C^\infty(M)^G \to C^\infty(G \times_{G_x} S_x)^G = C^\infty(S_x)^{G_x}$$

is open. Observe also that the restriction

$$C^\infty(M)^G \to C^\infty(M^G)$$

is open. Thus if $f \in C^\infty(M)^G$ and $f \in U$ with U open, then there exists $f_0 \in U$ such that $f_0|M^G$ is a Morse function.

Let $p_1, p_2, \ldots, p_k$ be the critical points of $f_0|M^G$ with pairwise disjoint disc neighbourhoods $(5/2)D(W_i)$ in M. Using the following lemma it is possible to approximate f_0 by a function f_1, with no degenerate critical points on $\cup D(W_i)$.

1.8.8 LEMMA. *Suppose that $f:W \to R$ is an invariant smooth function such that $f|W^G$ has only 0 as a critical point and 0 is non-degenerate for $f|W^G$. Then there exists an invariant smooth function $f' = W \to R$ such that $f' = f$ off $2D(W)$, $f'|D(W)$ has no degenerate critical points and f' is an arbitrarily close approximation to f.*

PROOF. Let Q be orthogonal projection onto $(W^G)^\perp$ the orthogonal complement of W^G in W. Let $\lambda: R \to I$ be a smooth decreasing function such that $\lambda = 1$ if $t \leq 1/2$ and $\lambda = 0$ if $t \geq 1$. Define f' by

$$f'(w) = f(w) + \alpha\lambda(||(2w/3)||^2)||Q(w)||^2.$$

Then $f' = f$ on W^G and $f' = f$ on $(3/2)D(W)$, proving the first assertion. Also, $0 \in W$ is the only critical point on $D(W)$. Calculation of Hessian's yields

$$H(f') = \begin{pmatrix} H(f|W^G) & C \\ D & B+2\alpha I \end{pmatrix}$$

where B, C, D depend on f near 0. Since $f|W^G$ is non-degenerate, $H(f|W^G)$ is invertible and hence det $H(f')$ is a non-trivial polynomial in α (use row and column operations to eliminate C and D). The non-degeneracy and approximation properties follow by choosing α to be small and not a root of this polynomial.

We now return to the proof of Theorem 1.8.7. Use Lemma 1.8.8 to get an approximation f_1 to f_0 that is non-degenerate on M^G and hence non-degenerate on a closed tube $M^G \subseteq E \subseteq M$. The construction can be done so that f_1 has no critical points on some closed collar D of ∂M.

By the inductive hypothesis $ND_G(G \times_{G_x} S_x, M)$ is dense if $x \in \text{Int}(M)$ and $G_x \neq G$. Let U_1 be given by

$$U_1 = U \cap ND_G(E,M) \cap ND_G(D,M)$$

so that elements of U_1 have no critical points on D if U is small enough. Notice that $f \in U_1$ so that $U_1 \neq \emptyset$. It is sufficient now to show that $ND_G(M,M) \cap U_1 \neq \emptyset$. To see this choose slices S_i such that the interiors of $G \times_{G_i} S_i$ cover $\overline{M-(E \cup D)}$. Then each $ND_G(G \times_{G_i} S_i, M)$ is open dense which implies that the finite intersection

$$W = \bigcap_i ND_G(G \times_{G_i} S_i, M)$$

is also open dense which in turn implies that

$$U_1 \cap W \neq \emptyset .$$

But it is immediate that

$$U_1 \cap W \subseteq U \cap \overline{ND_G(M,M)}$$

since function in W have no critical points in $\overline{M-(E \cup D)}$ while those in U_1 have none in $E \cup D$.

1.8.9 COROLLARY. *Invariant Morse functions exist.*

PROOF. Use a collar neighbourhood of ∂M to get a smooth G-invariant function, constant on every component of ∂M and with no critical points on ∂M. All the approximations of Theorem 1.8.7 work for such functions.

Note that generally our Morse functions have $f(M) \subseteq I$, $f(\partial_0 M) = 0$, $f(\partial_1 M) = 1$ where $\partial_0 M + \partial_1 M = \partial M$.

1.8.10 DEFINITION. A G invariant vector field X is *gradient-like* for a G invariant Morse function if $Xf > 0$ at all non-critical points, and if for a critical point p having a slice S_p on which

$$f(x) = |P(x)|^2 - |(I-P)(x)|^2$$

then on this slice X has the form

$$X(q) = P(q) - (I-P)(q).$$

1.8.11 LEMMA. *Every invariant Morse function admits an invariant gradient-like vector field.*

PROOF. Construct an invariant metric on M such that at critical orbits the conditions in the Morse lemma are satisfied. Let $X = \text{grad}(f)$, the gradient of f.

1.8.12 COROLLARY. *Let $f: M \to I$ be an invariant smooth function with no critical points. Then M is equivariantly diffeomorphic to $\partial_0 M \times I$, furthermore if $\phi: M \to \partial_0 M \times I$ is the diffeomorphism then $\phi | \partial_0 M : \partial_0 M \to \partial_0 M \times \{0\}$ is the identity.*

The proof is exactly the same as in the non-equivariant case. (See for example [Hirsch] etc.)

A closely related idea to that of a surgery is that of *equivariantly attaching an (H, V_1, V_2) handle*. To define this suppose that M is a G manifold with boundary $\partial_- M + \partial_+ M$. Suppose that V_1, V_2 are both H modules and that

$$\dim M = \dim G - \dim H + \dim V_1 + \dim V_2.$$

Let $\phi: G \times_H (S(V_1) \times D(V_2)) \to \partial_+ M$ be an equivariant regular embedding. The G manifold formed by equivariantly attaching an (H, V_1, V_2) handle along ϕ is

$$N = M \cup_{\partial_+ M} \partial_+ M \times I \cup_\phi G \times_H D(V_1) \times D(V_2).$$

This is a G manifold with boundary

$$\partial_- M + (\partial_+ M \times \{1\} - \phi(G \times_H (S(V_1) \times D(V_2))) \times \{1\}) \cup_\phi G \times_H D(V_1) \times S(V_2).$$

If M and M' are closed G manifolds and if M' results from M by a surgery of type $[H, V_1, V_2]$ then the trace W of the surgery is the result of equivariantly attaching an (H, V_1, V_2) handle to $M \times I$.

We can now prove exactly as in the non-equivariant case the existence of *equivariant handle decompositions*.

1.8.13 THEOREM. *Suppose M is a G manifold with boundary $\partial_- M + \partial_+ M$. Then there is a finite sequence of invariant smooth submanifolds with the same dimension as M:*

$$\partial_- M \times [0, \alpha] = Q_0 \subseteq Q_1 \subseteq \ldots \subseteq Q_r = M$$

such that $Q_i \subseteq Interior(Q_{i+1})$, (with $\partial_- M$ removed), and Q_{i+1} is obtained from Q_i by smoothly attaching a finite number of disjoint G handles along $\partial_+ Q_i$.

Theorem 1.8.1 now follows. We note that a similar result to Theorem 1.8.1 also holds for the singular bordism theory $N_*(X)$, with surgery of singular manifolds defined in an obvious way.

1.9 HISTORICAL NOTE

P.E. Conner and E.E. Floyd introduced the notion of bordism with "families of subgroups" in [Conner and Floyd, 2]. These ideas were subsequently used by many authors, for example [Stong, 2]. The idea of using "families of slice types" seems to appear first in the book [Karras, Kreck, Neumann and Ossa; Chapter 3]. The first place where I came across this notion was in a paper of R.E. Stong [Stong, 5]. It was this paper which was instrumental in my interest in these families and in the subsequent applications. The reader may wish to look at [Pulikowski, 1,2] and [Kosniowski, 1] for some related ideas.

The results of section 1.5 are essentially contained in [Conner and Floyd, 2].

The existence of invariant Morse functions etc., was proved by

A.G. Wassermann in [Wassermann]. The proof presented here is due to R.E. Schultz and I am indebted to him for permission to use his notes. The proof follows that given for the non-equivariant case in [Lang]. Another approach to prove the smooth G handle decomposition theorem has been suggested by E. Bierstone in [Bierstone]. A good exposition of this may be found in [Lashof and Rothenberg].

2 Cutting and pasting groups

2.1 THE SK GROUPS

Let M be a closed manifold and let $L \subset M$ be a closed submanifold of codimension 1 with trivial normal bundle. If we cut M open along L we obtain a manifold M' with boundary $\partial M' = L+L$ (disjoint union). By pasting these two copies of L together again in a different way we obtain a new closed manifold M_1. We say that M_1 has been obtained from M by cutting and pasting. If N has been obtained from M by sequence of cuttings and pastings then we say that M and N are SK equivalent (Sneiden and Kleben). We shall make the assumption that L separates M - this is no loss of generality since if L does not separate M then the union of L with a second copy of L, suitably embedded near L, will separate M. Therefore we can write

$$M = N \cup_\phi N', \quad M_1 = N \cup_\psi N'$$

where $\phi, \psi : \partial N \to \partial N'$ are diffeomorphisms.

More generally, let X be a space. We say that the singular n manifolds (M_1, f_1) and (M_2, f_2) are obtainable from each other by cutting and pasting if

(i) M_2 has been obtained from M_1 by cutting and pasting, in particular

$$M_1 = N \cup_\phi N', \quad M_2 = N \cup_\psi N'$$

for some diffeomorphisms $\phi, \psi : \partial N \to \partial N'$, and

(ii) there are homotopies

$$f_1|N \simeq f_2|N, \quad f_1|N' \simeq f_2|N'.$$

Two singular n manifolds $(M_1, f_1), (M_2, f_2)$ are said to be SK equivalent if (M_2, f_2) can be obtained from (M_1, f_1) by a sequence of cuttings and pastings in X. It is easy to see that SK equivalence is an equivalence relation. Disjoint union makes the set of equivalence classes into an abelian semigroup. The Grothendieck group of SK equivalence classes of singular n manifolds in X is denoted by $SK_n(X)$. We shall write elements of $SK_n(X)$ as $[M, f], [M', f']$ etc. where $[M, f]$ is the equivalence class containing (M, f). If X is a point then we write $SK_n(pt)$ simply as SK_n, elements of which are denoted by $[M]$ etc.

By defining

$$SK_* = \bigoplus_{n \geq 0} SK_n$$

we get a graded ring, using cartesian product of manifolds as multiplication. Similiarly

$$SK_*(X) = \bigoplus_{n \geq 0} SK_n(X)$$

becomes a graded SK_* module. The ring SK_* is isomorphic to the integral polynomial ring on one generator of dimension 2 which may be taken to be RP^2. A proof of this fact will appear in section 2.3.

2.2 EQUIVARIANT SK GROUPS

We define the equivariant SK group SK_n^G as the Grothendieck group of n dimensional G manifolds modulo the relations given by equivariant cuttings and pastings. The submanifolds $L \subset M$ that we are allowed to cut open along are those satisfying

(i) L is a G invariant codimension 1 submanifold of M,

(ii) L has trivial normal bundle in M, and

(iii) the normal bundle of L in M is equivariantly equivalent to $L \times R$ with trivial action of G on R.

After cutting open along L we obtain a manifold M' with boundary. Condition (i) ensures that M' is a G manifold, condition (ii) ensures that $\partial M' = L+L$ while condition (iii) ensures that each of the two copies of L are G invariant in M'. As in the non-equivariant case it is no loss of generality to insist that L separates M. By pasting these two copies of L together via some other equivariant diffeomorphism we obtain a closed G manifold M_1. We say that M_1 has been obtained from M by equivariant cutting and pasting. If M_1 has been obtained from M by a sequence of equivariant cuttings and pastings then we say that M_1 and M are SK^G equivalent. This is an equivalence relation on the set of n dimensional G manifolds. The equivalence classes form an abelian semigroup if we use disjoint union as addition. The Grothendieck group of this semigroup is then denoted by SK_n^G. The equivalence class containing the G manifold M is denoted by $[M]$.

By defining

$$SK_*^G = \bigoplus_{n \geq 0} SK_n^G$$

we get an SK_* module with multiplication given by cartesian product of manifolds. One of the objects of this book is to describe the SK_* module SK_*^G.

As in bordism, SK_*^G can be generalised by using families of

G slice types. Let F be a family of G slice types. We denote by $SK_n^G[F]$ the SK group resulting from equivariant cuttings and pastings of n dimensional G manifolds of type F. There is an obvious SK_* module structure on

$$SK_*^G[F] = \bigoplus_{n \geq 0} SK_n^G[F] .$$

If $F' \subseteq F$ is a pair of families of G slice types then there is an obvious SK_* homomorphism

$$i: SK_*^G[F'] \to SK_*^G[F]$$

which is clearly injective - just check the definition of equivariant cutting and pasting of G manifolds of type F' and F.

We shall refrain from defining the equivariant SK groups for pairs of families of G slice types except, indirectly, for the special case which leads to SK groups of vector bundles. Let $[H;U]$ be a G slice type. We shall denote by $SK_n^G[H;U]$ the SK group resulting from equivariant cuttings and pastings of G vector bundles of type $[H;U]$ over closed manifolds (here n denotes the total dimension of the vector bundle). By equivariant cutting and pasting of G vector bundles we mean the following: Suppose that E is a G vector bundle over M and suppose that L is a G submanifold of M satisfying the conditions (i), (ii), and (iii) given at the beginning of this section. Cutting along $E|L$ gives a G vector bundle E' over M' (which is a G manifold with boundary $\partial M' = L_1 + L_2$, $L_1 = L_2 = L$). We now paste via an equivariant diffeomorphism

$$\phi: E'|L_1 \to E'|L_2$$

which is a vector space isomorphism fibrewise. There results a

G vector bundle E_1 over some manifold M_1. In fact if $M = N \cup_1 N'$ then $M_1 = N \cup_\phi N'$ and $E = E|N \cup E|N'$ while $E_1 = E|N \cup_\phi E|N'$. If E is a G vector bundle of type $[H;U]$ then so is E_1. Note that from section 1.5 we have

$$SK_*^G[H;U] \cong SK_{*-k}(B\Gamma(H;U))$$

where $k = \dim U$.

2.2.1 EXAMPLE. Consider the Möbius strip and the cylinder as $\mathbb{Z}/2$ vector bundles of type $[\mathbb{Z}/2, \tilde{R}]$ where $\tilde{R}$ denotes the $\mathbb{Z}/2$ module with underlying space R and $\mathbb{Z}/2$ action given by multiplication by $\{\pm 1\}$.

 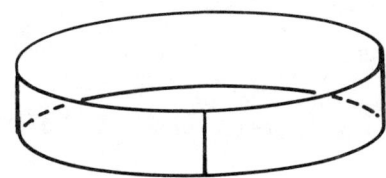

Cutting and pasting along the lines indicated shows that they are $SK^{\mathbb{Z}/2}$ equivalent as $\mathbb{Z}/2$ vector bundles.

Suppose that $F' \subseteq F$ is a pair of families of G slice types such that $F - F' = \{[H;U]\}$ then there is an SK_* module homomorphism

$$\nu_{[H;U]} : SK_*^G[F] \to SK_*^G[H;U] \ .$$

This is defined by assigning to a G manifold M the normal bundle of the G submanifold N consisting of those points in M with slice type precisely $[H;U]$. We now have the following sequence

51

of SK_* module homomorphisms

$$0 \to SK_*^G[F'] \xrightarrow{i} SK_*^G[F] \xrightarrow{\nu} SK_*^G[H;U]$$

with i injective and $\nu i = 0$ clearly. It is not easy, in general, to show that this sequence is exact. The order of difficulty depends on the slice type $[H;U]$ and the families $F' \subseteq F$.

2.3 BORDISM AND SK EQUIVALENCE

In this section we shall consider how bordant G manifolds differ under SK^G equivalence.

2.3.1 THEOREM. *Suppose that M,M' are G manifolds of type F, where F is a family of G slice types. If M' results from M by a surgery of type $[H,V_1,V_2;F]$ then*

$$[M] + [G \times_H S(V_1 \times V_2)]$$
$$= [M'] + [G \times_H (S(V_1) \times S(V_2 \times R))]$$

in $SK_*^G[F]$.

PROOF. Let

$$\phi: G \times_H (S(V_1) \times D(V_2)) \to M$$

be the embedding on which surgery was done. We have

$$M + G \times_H S(V_1 \times V_2) = (M - \phi(G \times_H (S(V_1) \times D(V_2))))$$
$$\cup\; G \times_H (S(V_1) \times D(V_2))$$
$$+\; G \times_H (S(V_1) \times D(V_2)) \cup G \times_H (D(V_1) \times S(V_2))$$

and

$$M' + G \times_H (S(V_1) \times S(V_2 \times R)) =$$

$$M - \phi(G \times_H S(V_1) \times D(V_2)) \;\cup\; G \times_H (D(V_1) \times S(V_2))$$

$$+ \; G \times_H (S(V_1) \times D(V_2)) \;\cup\; G \times_H (S(V_1) \times D(V_2))$$

with obvious identification on the boundaries. Thus we can see that

$$M' + G \times_H (S(V_1) \times S(V_2 \times R))$$

results from

$$M + G \times_H S(V_1 \times V_2)$$

by equivariant cutting and pasting.

In section 1.8 we showed that

$$G \times_H (S(V_1 \times R) \times S(V_2))$$

resulted from

$$G \times_H S(V_1 \times V_2)$$

by a surgery of type $[H, V_1, V_2; F]$ where F is some family containing the slice types contained in the G manifold $G \times_H D(V_1 \times V_2)$. Hence by using Theorem 2.3.1 we obtain

$$[G \times_H S(V_1 \times V_2)] + [G \times_H S(V_1 \times V_2)]$$
$$= [G \times_H (S(V_1 \times R) \times S(V_2))] + [G \times_H (S(V_1) \times S(V_2 \times R))]$$

or

$$[G \times_H (S(V_1 \times R) \times S(V_2))]$$
$$= 2[G \times_H S(V_1 \times V_2)] - [G \times_H (S(V_1) \times S(V_2 \times R))].$$

2.3.2 COROLLARY. *Let M and M' be n dimensional G manifolds of type F. If M and M' are F bordant then*

$$[M] = [M'] + \sum a(H, U_1, U_2)[G \times_H (S(U_1) \times S(U_2))]$$
$$+ \sum b(H, U)[G \times_H S(U)]$$

for some integers $a(H, U_1, U_2)$, $b(H, U)$. The first sum is taken over all subgroups $H \subseteq G$ and all H modules U_1, U_2 satisfying

(i) $\dim U_1 + \dim U_2 = n+1$,

(ii) $(U_1)^H = \{0\}$, and

(iii) $G \times_H D(U_1 \times U_2)$ is of type F.

The second sum is taken over all subgroups $H \subseteq G$ and all H modules U of dimension $n+1$ which satisfy the condition that $G \times_H D(U)$ is of type F.

PROOF. If M and M' are F bordant then we can find a sequence of F bordant G manifolds

$$M = M_0, M_1, \ldots, M_r = M'$$

such that M_i results from M_{i-1} by a surgery of type $[H_i, V_{i,1}, V_{i,2}; F]$. The result is proved by induction on r. We shall prove the case $r = 1$, so M' results from M by a surgery of type $[H, V_1, V_2; F]$. Theorem 2.3.1 tells us that

$$[M] = [M'] + [G \times_H (S(V_1) \times S(V_2 \times R))] - [G \times_H S(V_1 \times V_2)].$$

If $(V_1)^H \neq \{0\}$ then $V_1 = W_1 \times R$ as an H module so that

$$[G \times_H (S(V_1) \times S(V_2 \times R))] = [G \times_H (S(W_1 \times R) \times S(V_2 \times R))]$$
$$= 2[G \times_H S(W_1 \times V_2 \times R)]$$
$$- [G \times_H (S(W_1) \times S(V_2 \times R \times R))]$$

(by the equation above 3.3.2)

$$= 2[G \times_H S(V_1 \times V_2)] -$$

$$- \left[G \times_H (S(W_1) \times S(V_2 \times R^2))\right].$$

Thus we have

$$[M] = [M'] - \left[G \times_H (S(W_1) \times S(V_2 \times R^2))\right] + \left[G \times_H S(V_1 \times V_2)\right].$$

Continuing in this manner we obtain

$$[M] = [M'] + (-1)^m \left[G \times_H (S(W_m) \times S(V_2 \times R^{m+1}))\right]$$
$$- (-1)^m \left[G \times_H S(V_1 \times V_2)\right]$$

where $V_1 = W_m \times R^m$. This process continues until $(W_m)^H = \{0\}$, the assertion then follows.

If G is the trivial group then the above corollary shows that if the n dimensional manifolds M and M' are bordant to each other then

$$[M] = [M'] + k[S^n]$$

for some integer k. We leave it to the reader to prove the following result which generalises the result just mentioned to the case of singular manifolds in X. (Alternatively, see [Karras, Kreck, Neumann and Ossa].)

2.3.3 LEMMA. *Suppose that (M,f) and (M',f') are singular n manifolds in a path connected space X. If (M,f) and (M',f') are bordant then*

$$[M,f] = [M',f'] + k[S^n, *]$$

where k is some integer and $*$ denotes the (unique up to homotopy) constant map $S^n \to X$.

Recall that the euler characteristic χ has the additivity property

$$\chi(X \cup Y) = \chi(X) + \chi(Y) - \chi(X \cup Y)$$

(at least for "nice" spaces, e.g., X, Y CW subspaces of $X \cup Y$ etc.) and recall also that the euler characteristic of a closed odd dimensional manifold is zero. (See for example [Dold, 1; Chapter V5]). Thus if

$$M = N \cup_\phi N', \quad M' = N \cup_\psi N'$$

for some ϕ, ψ then $\chi(M) = \chi(M')$. More generally if M and M' are SK equivalent then $\chi(M) = \chi(M')$.

2.3.4 COROLLARY. *If $(M,f) \sim (M',f')$ in $N_{2n}(X)$ then*

$$[M,f] = [M',f'] + (1/2)(\chi(M)-\chi(M'))[S^{2n}, *].$$

PROOF. Look at the euler characteristics of the manifolds in Lemma 2.3.3:

$$\chi(M) = \begin{cases} \chi(M') + 2k & \text{if } n \text{ is even} \\ \chi(M') & \text{if } n \text{ is odd.} \end{cases}$$

2.3.5 EXERCISE. Prove that the result in Corollary 2.3.4 is also true in the odd dimensional case.

2.4 FIBRATIONS AND SK EQUIVALENCE

In this section we shall investigate the SK equivalence of fibrations. The results will be crucial later on.

2.4.1 THEOREM. (i) $[S^1] = 0$ where S^1 has no G action.

(ii) Suppose M is an n dimensional G manifold of type F and that V is some H module, $H \subseteq G$. If M fibres equivariantly over $G \times_H S(V \times R)$ with typical fibre F then

$$[M] = [G \times_H (S(V \times R) \times F)]$$

in $SK_n^G[F]$. If furthermore F is a G manifold then

$$[M] = [G \times_H S(V \times R)][F].$$

(iii) Suppose M is an n dimensional G manifold of type F and that $V = V_1 \times V_2 \times \ldots \times V_r$ is a product of one dimensional H modules, $H \subseteq G$. If M fibres equivariantly over $G \times_H RP(V)$ with typical fibre F then

$$[M] = [G \times_H (RP(V) \times F)].$$

If furthermore F is a G manifold then

$$[M] = [G \times_H RP(V)][F].$$

(iv) The result in (iii) also holds if we assume that each V_i is a one dimensional complex H module and if we replace $G \times_H RP(V)$ by $G \times_H CP(V)$.

(v) The above results (ii), (iii) and (iv) also hold in the non-equivariant singular case, i.e., in $SK_*(X)$.

PROOF. (i) Let $N = N' = I + I$. We can paste N to N' in two ways to obtain either S^1 or $S^1 + S^1$

Hence $[S^1] = 2[S^1]$ or $[S^1] = 0$.

(ii) Write $G \times_H S(V \times R)$ as

$$G \times_H S(V \times R) = G \times_H D(V) \cup G \times_H D(V)$$

with the obvious identifications in the boundary $G \times_H S(V)$. Any equivariant fibration over $G \times_H D(V)$ is of the form $G \times_H (D(V) \times F)$ since $D(V)$ is H contractible (V is an H module) and equivariant fibrations over G/H are of the form $G \times_H F$. Thus we can write

$$M = G \times_H (D(V) \times F) \cup_\phi G \times_H (D(V) \times F)$$

for some suitable ϕ. After cutting and pasting we can obtain $G \times_H (S(V \times R)) \times F)$. If F is also a G manifold then $G \times_H F \cong (G/H) \times F$ and

$$G \times_H (S(V \times R) \times F) \cong (G \times_H S(V \times R)) \times F$$

via the map

$$(g, x, y) \to (g, x, gy)$$

for $g \in G$, $x \in S(V \times R)$, $y \in F$.

(iii) The proof is by induction on $r = \dim V$, with the case $r = 1$ being trivial. Assume that $r > 1$, and recall that we are assuming that $V = V_1 \times V_2 \times \ldots \times V_r$ is the product of one dimensional H modules. We have

$$G \times_H RP(V) = G \times_H RP(V \otimes V_r)$$
$$= G \times_H RP(U \times R)$$

where $U = (V_1 \times \ldots \times V_{r-1}) \otimes V_r$. We now write $G \times_H RP(U \times R)$ as

$$G \times_H RP(U \times R) = G \times_H D(U) \cup G \times_H N$$

where N is diffeomorphic to the equivariant normal disc bundle of $RP(U)$ in $RP(U\times R)$. Observe that the fibre F is an H manifold and define M_0, M_1 as follows

$$M_0 = M|G \times_H D(U) \quad + \quad G \times_H (N\times F)$$
$$= G \times_H (D(U)\times F) + \quad G \times_H (N\times F)$$
$$M_1 = M|N \quad + \quad G \times_H (D(U)\times F) \ .$$

By pasting M_0 to M_1 in two ways we obtain

$$M_0 \cup_\phi M_1 = M + \quad G \times_H (RP(U\times R)\times F)$$
$$M_0 \cup_\psi M_1 = L + \quad G \times_H (S(U\times R)\times F)$$

where ϕ pasted the first, second part of M_0 to the first, second part of M_1 respectively while ψ pasted the first, second part of M_0 to the second, first part of M_1 respectively. The manifold L is an equivariant fibration over the double $D(N) = N \cup N$ of N with fibre F. However $D(N)$ fibres over $G \times_H RP(U)$ with fibre S^1. Thus L is an equivariant fibration over $G \times_H RP(U)$ with fibre F' where F' is a fibration over S^1 with fibre F. Hence by the inductive hypothesis and part (ii) we have

$$[L] = [G \times_H (RP(U)\times F')]$$
$$= [G \times_H (RP(U)\times F\times S^1)]$$
$$= [G \times_H (RP(U)\times F)][S^1] \quad \text{(since } S^1 \text{ has trivial action)}$$
$$= 0 \quad \text{(by part (i))}.$$

Thus we have

$$[M] + [G \times_H (RP(U\times R)\times F)] = [G \times_H (S(U\times R)\times F)].$$

By putting $M = G \times_H (RP(U\times R)\times F)$ in this equality we obtain

$$2[G \times_H (RP(U\times R)\times F)] \cong [G \times_H (S(U\times R)\times F)]$$

and so

$$[M] = [G \times_H (RP(U \times R) \times F)].$$

Finally if F is also a G manifold we have

$$G \times_H (RP(U \times R) \times F) \cong (G \times_H RP(U \times R)) \times F.$$

(iv) The proof is analogous to that in (iii).

(v) The proof is left for the reader, the essential details have been given above.

2.5 CALCULATION OF SK_*

The calculation of SK_* proceeds by using results from the preceding two sections and knowledge of N_* (section 1.1).

2.5.1 THEOREM. (i) $SK_* \cong \mathbb{Z}[RP^2]$.

(ii) The isomorphism

$$SK_n \to \begin{cases} \mathbb{Z} & \text{if } n \text{ is even} \\ 0 & \text{if } n \text{ is odd} \end{cases}$$

is given by the euler characteristic.

To prove the result let us first look at the generators of N_*. These are x_i of dimension i with i not of the form 2^s-1. The generators x_{2i} may be taken to be RP^{2i}. The generators x_{2i-1} may be the Dold manifold $P(m,n)$ where $m = 2^r-1$, $n = 2^r s$ and $2i = 2^r(2s+1)$. The Dold manifolds are SK equivalent to 0 because they are fibrations over CP^n with fibre S^m, m odd. But S^m, m odd is a fibration over $CP^{(m-1)/2}$ with fibre S^1 and hence SK equivalent to 0. The next result will tell us what the even dimensional generators are SK equivalent to.

2.5.2 LEMMA. *If M is a $\mathbb{Z}/2$ manifold and E denotes the normal bundle to the fixed point set in M then M and $RP(R \times E)$ are bordant.*

To prove this let $\Gamma(M) = S^1 \times_{\mathbb{Z}/2} M$ where $\mathbb{Z}/2$ acts antipodally on S^1. Let $\mathbb{Z}/2$ act on $\Gamma(M)$ by $(t,m) \to (\bar{t},m)$ and let N denote an equivariant open tubular neighborhood of the fixed point set in $\Gamma(M)$. Thus $\Gamma(M)-N$ has a free involution. An easy check reveals that the quotient $(\Gamma(M)-N)/(\mathbb{Z}/2)$ is a manifold with boundary $M + RP(R \times E)$.

On RP^{2k} ($k > 1$) consider the following $\mathbb{Z}/2$ action
$$[x_0, x_1, \ldots, x_{2k}] \to [-x_0, -x_1, x_2, \ldots, x_{2k}].$$
The fixed point set is $RP^1 + RP^{2k-2}$. Let E_{2k-1}, E_2 denote the normal bundles of RP^1, RP^{2k-2} respectively in RP^{2k}. So
$$RP^{2k} \sim RP(R \times E_{2k-1}) + RP(R \times E_2).$$
The manifold $RP(R \times E_{2k-1})$ is a fibration over $RP^1 = S^1$ and hence is SK equivalent to 0. The manifold $RP(R \times E_2)$ is a fibration over RP^{2k-2} with fibre RP^2, hence is SK equivalent to $RP^{2k-2} \times RP^2$. Using induction and Corollary 2.3.4 we deduce that
$$[RP^{2k}] = [RP^2]^k + \alpha_k [S^{2k}]$$
where $\alpha_k = \frac{1}{2}(\chi(RP^{2k}) - \chi((RP^2)^k)) = 0$. Thus $[RP^{2k}] = [RP^2]^k$.

Suppose now that M is a closed n dimensional manifold. We know that M is bordant to a manifold M' which can be expressed as a homogeneous polynomial of degree n in the generators x_i of N_*. From the above calculations we see that any homogeneous polynomial in the x_i is SK equivalent to a multiple of $[RP^2]^k$ if $n = 2k$ or to 0 if n is odd. By Corollary 2.3.4 we therefore have

$$[M] = \begin{cases} \alpha_M [RP^2]^k + \beta_M [S^{2k}] & \text{if } n = 2k \\ \beta_M [S^{2k+1}] & \text{if } n = 2k+1 \end{cases}$$

for some integers α_M, β_M.

Since S^n is a double covering of RP^n we have by 2.4.1(iii)

$$[S^n] = 2[RP^n].$$

Thus in the case $n = 2k$ we have

$$[S^{2k}] = 2[RP^2]^k.$$

In the case $n = 2k+1$ we have a fibration $S^{2k+1} \to CP^k$ with fibre S^1 and so by 2.4.1(iv) and 2.4.1(i) we have

$$[S^{2k+1}] = 0.$$

In conclusion we have

$$[M] = \begin{cases} (\alpha_M + 2\beta_M)[RP^2]^k & \text{if } n = 2k \\ 0 & \text{if } n = 2k+1 \end{cases}$$

and the theorem follows.

2.5.3. EXERCISE. Reprove Theorem 2.5.1 using the Milnor manifolds $H_{m,n}$ (section 1.1) as the odd dimensional generators of N_*.

2.6 SPLITTING HOMOMORPHISMS

In section 2.2 we described a sequence

$$0 \to SK_*^G[F'] \xrightarrow{i} SK_*^G[F] \xrightarrow{\nu} SK_*^G[H;U]$$

where $F' \subseteq F$ is a pair of families of G slice types with $F - F' = \{[H;U]\}$. We know that i is injective and that $\nu i = 0$.

In this section we shall give some conditions on the families F',F in order that the sequence be exact. In fact this section is the SK analogue of section 1.7.

2.6.1 THEOREM. *Let* $F' \subseteq F$ *be families of G slice types with* $F-F' = \{[H;U]\}$. *If* M_1 *and* M_2 *are G manifolds of type F with* $\nu(M_1) = \nu(M_2)$ *where* $\nu = \nu_{[H;U]}$, *then* $[M_1]+[M_2'] = [M_2]+[M_2']$ *where* M_1' *and* M_2' *are G manifolds of type* F'.

PROOF. Denote by N_i, $i = 1,2$, a submanifold of M_i which is diffeomorphic to the unit (closed) disc bundle $D\nu(M_i)$ of $\nu(M_i)$. (In other words N_i is a closed G invariant unit disc tubular neighbourhood of the set of points in M_i with slice type precisely $[H;U]$.) The submanifold ∂N_i of M_i satisfies the conditions (i), (ii) and (iii) of section 2.2 and hence we may cut M open along ∂N_i so that we have

$$M_i = (M_i - \overset{o}{N}_i) \cup N_i .$$

Since $\nu(M_1) = \nu(M_2)$ we may identify N_1 with N_2. Define manifolds M_1' and M_2' by

$$M_1' = (M_1 - \overset{o}{N}_1) \cup (M_2 - \overset{o}{N}_2)$$
$$M_2' = (M_1 - \overset{o}{N}_1) \cup (M_1 - \overset{o}{N}_1) .$$

It is clear that M_1' and M_2' are G manifolds of type F'. It is easy to see that

$$[M_1]+[M_1'] = [M_2]+[M_2']$$

after performing some elementary cutting and pasting.

2.6.2 COROLLARY. *Let $F' \subseteq F$ be families of G slice types with $F - F' = \{[H;U]\}$. Suppose that for all G vector bundles E of type $[H;U]$ we can assign a G manifold $q(E)$ of type F such that*

(i) $\nu_{[H;U]} q(E) = E$, *and*

(ii) *if* $[E] = [E']$ *then* $[q(E)] = [q(E')]$.

Then the following sequence

$$0 \to SK_*^G[F'] \xrightarrow{i} SK_*^G[F] \xrightarrow{\nu} SK_*^G[H;U] \to 0$$

is a short split exact sequence.

PROOF. We know that i is injective and that $\nu i = 0$. Condition (i) ensures that ν is surjective and the second condition ensures that q defines an SK_* homomorphism $q : SK_*^G[H;U] \to SK_*^G[F]$. It remains to show that $\ker(\nu) \subseteq \operatorname{im}(i)$.

Suppose that $[M_1] - [M_2] \in SK_*^G[F]$ with $\nu([M_1] - [M_2]) = 0$. Since $\nu(M_i) = \nu(q\nu(M_i))$ for $i = 1,2$ we have that

$$[M_i] + [M_i'] = [q\nu(M_i)] + [M_i'']$$

for some G manifolds M_i', M_i'' of type F'. But $[\nu(M_1)] = [\nu(M_2)]$ implies that $[q\nu(M_1)] = [q\nu(M_2)]$ and so we deduce that

$$[M_1] - [M_2] = [M_1''] - [M_2''] - [M_1'] + [M_2']$$

showing that $[M_1] - [M_2]$ is in the image of i.

2.6.3 COROLLARY. *Let $F' \subseteq F$ be families of G slice types with $F - F' = \{[H;U]\}$. If*

(i) $G \times_H RP(R \times U)$ *is a G manifold of type F, and if*

(ii) *there does not exist a subgroup K of H of index 2 such that $U = V_K \oplus W$ with $V_K \not\subseteq W$ and $V_K \otimes W = W$,* *then the sequence*

$$0 \to SK_*^G[F'] \to SK_*^G[F] \to SK_*^G[H;U] \to 0$$

is a short split exact sequence.

PROOF. Define $q(E)$ by

$$q(E) = RP(R \times E).$$

Condition (i) ensures that $q(E)$ is a G manifold of type F. Condition (ii) ensures, by Theorem 2.7.1, that $\nu_{[H;U]} q(E) = E$. Finally, if $E = E_1 \cup_\phi E_2$ then

$$RP(R \times E) = RP(R \times E_1) \cup_\psi RP(R \times E_2)$$

where ψ is the map induced by ϕ. Thus if $[E] = [E']$ then $[q(E)] = [q(E')]$ and the result follows from Corollary 2.6.2.

The conditions in the above corollary always hold in the case that G is a group of odd order. We immediately get the next result

2.6.4 COROLLARY. *If G is a finite abelian group of odd order then*

$$SK_*^G \cong \bigoplus_{[H;U] \,\varepsilon\, St(G)} SK_*^G[H;U] \ .$$

Alternatively, we could write this as

$$SK_*^G \cong \bigoplus_{[H;U] \,\varepsilon\, St(G)} SK_{*-|U|}(B\Gamma(H;U)) \ .$$

Later on we shall in fact show that

$$SK_*(B\Gamma(H;U)) \cong SK_* \ .$$

2.7 BORDISM AND SPLITTINGS.

In this section we show how in proving exactness of the SK_*^G sequence bordism enters the scene. Let $F' \subseteq F$ be a pair of families of G slice types with $F - F' = \{[H;U]\}$.

2.7.1 THEOREM. *If M and M_0 are n dimensional G manifolds of type F with $\nu(M) \sim \nu(M_0)$ in $N_n^G[H;U]$ then*

$$[M] = [M_0] + [M'] - [M''] + \alpha[N][G \times_H S(R \times U)]$$

where M' and M'' are G manifolds of type F' while N is an (n-dim U) dimensional manifold with trivial G action and $\alpha = \pm 1$.

PROOF. Since $\nu(M) \sim \nu(M_0)$ we have that $M \sim M_0 + M_1$ (in $N_n^G[F]$) where M_1 is a G manifold of type F'. (The sequence

$$\cdots \to N_n^G[F'] \xrightarrow{i} N_n^G[F] \xrightarrow{\nu} N_n^G[H;U] \to$$

is exact.) By Corollary 2.3.2 we have

$$[M] = [M_0] + [M_1] + \sum a(K, V_1, V_2)[G \times_K (S(V_1) \times S(V_2))]$$
$$+ \sum b(K, V)[G \times_K S(V)]$$

where (i) $K \subseteq G$

(ii) $\dim(V_1) + \dim(V_2) = n+1 = \dim V$

(iii) $(V_1)^K = \{0\}$

(iv) $G \times_K D(V_1 \times V_2)$ is of type F, and

(v) $G \times_K D(V)$ is of type F.

Now $F - F' = \{[H;U]\}$; so if $G \times_K D(V)$ is of type F then $G \times_K D(V)$ must also be of type F' except in the case $G \times_H D(U \times R^\ell)$ where $\ell = n+1 - \dim U$. Thus we see that

$$[M_1] + \sum a(K,V_1,V_2)[G \times_K (S(V_1) \times S(V_2))]$$

$$+ \sum_{[K;V] \neq [H;U \times R^\ell]} b(K;V)[G \times_K S(V)]$$

belongs to $SK_n^G[F']$. We write this element as $M_2 - M_3$ with M_2, M_3 being G manifolds of type F'. We therefore obtain

$$[M] = [M_0] + [M_2] - [M_3] + b[G \times_H S(U \times R^\ell)]$$

for some integer b.

Consider the G manifold P given by

$$P = (G \times_H (S(U \times R) \times S(R^\ell)))/(\mathbb{Z}/2)$$

where the $\mathbb{Z}/2$ action is that induced from multiplication by -1 on R and R^ℓ. We see that

$$\nu(P) = (G \times_H U) \times S(R^\ell)$$
$$= \nu(G \times_H S(U \times R^\ell)) .$$

From Theorem 2.6.1 we deduce that

$$[P] + [M_4] = [G \times_H S(U \times R^\ell)] + [M_5]$$

where M_4 and M_5 are of type F'. The G manifold P fibres equivariantly over $S(R^\ell)/(\mathbb{Z}/2) \cong RP^{\ell-1}$ with fibre $G \times_H S(R \times U)$ and so by Theorem 2.4.1(iii) we have

$$[P] = [RP^{\ell-1}][G \times_H S(R \times U)] .$$

We therefore obtain

$$[M] = [M_0] + [M_2] - [M_3] + b[M_4] - b[M_5] + b[RP^{\ell-1}][G \times_H S(R \times U)],$$

from which the result follows.

2.7.2 COROLLARY. *Suppose that M_1, M_2 are n dimensional G manifolds of type F with $[\nu(M_1)] = [\nu(M_2)]$ and $\nu(M_1) \sim \nu(M_2)$ then*

$$[M_1] + [M''] = [M_2] + [M']$$

where M' and M" are G manifolds of type F'.

PROOF. From the previous result we see that since $\nu(M_1) \sim \nu(M_2)$ we have

$$[M_1] = [M_2] + [M'] - [M''] + \alpha[N][G \times_H S(R \times U)] .$$

Because $[\nu(M_1)] = [\nu(M_2)]$ we must have

$$\alpha[N] \times 2 \times [G \times_H U] = 0 \text{ in } SK_n^G[H;U] .$$

But under the natural forgetful map

$$SK_n^G[H;U] \to SK_{\ell-1}$$

which just looks at the base space (ignoring group action) of the G vector bundle we have

$$2\alpha[N][G \times_H U] \to 2\alpha|G/H|[N]$$

where $|G/H|$ denotes the number of elements in $|G/H|$. Since $SK_{\ell-1}$ has no torsion (see section 2.5) and $2|G/H| \neq 0$ we have

$$\alpha[N] = 0$$

which means that

$$[M_1] = [M_2] + [M'] - [M'']$$

thus establishing the corollary.

2.7.3 COROLLARY. *If $F' \subseteq F$ and $F - F' = \{[H;U]\}$ then the sequence*

$$0 \to SK_*^G F' \otimes \mathbb{Z}[\tfrac{1}{2}] \xrightarrow{i} SK_*^G[F] \otimes \mathbb{Z}[\tfrac{1}{2}] \xrightarrow{\nu} SK_*^G[H;U] \otimes \mathbb{Z}[\tfrac{1}{2}] \to 0$$

is a short split exact sequence.

PROOF. This follows from 2.7.2 and the fact that $N_*^G[F]$ has only 2-torsion. A direct proof may be obtained (as in [Karras, Kreck, Neumann and Ossa]) by using the homomorphism

$$q: SK_*^G[H;U] \otimes \mathbb{Z}[\tfrac{1}{2}] \to SK_*^G[F] \otimes \mathbb{Z}[\tfrac{1}{2}]$$

$$E \otimes 1 \to DD(E) \otimes \tfrac{1}{2}$$

where $DD(E)$ denotes the double of the disc bundle of E.

2.7.4 COROLLARY.

$$SK_*^G \otimes \mathbb{Z}[\tfrac{1}{2}] \cong \bigoplus_{[H;U] \in St(G)} SK_*^G[H;U] \otimes \mathbb{Z}[\tfrac{1}{2}] \quad .$$

2.8 HISTORICAL NOTE.

The theory of SK invariants - invariants under cutting and pasting was started by K. Jänich in [Jänich, 1,2]. The non-equivariant results on SK_* come from the book [Karras, Kreck, Neumann and Ossa] although the proofs here are different. For more history of SK_* the reader is referred to the book [Karras, Kreck, Neumann and Ossa]. The reader is warned that in that book the authors use the notation SK_*^O where we use SK_*, their use of the symbol SK_* is for the oriented analogue.

Lemma 2.5.2 was originally proved in [Conner and Floyd, 1].

3 Classifying spaces

3.1 BORDISM AND COHOMOLOGY

The purpose of this chapter is to calculate $N_*^G[H;U]$ and $SK_*^G[H;U]$ or equivalently $N_*(B\Gamma(H;U))$ and $SK_*(B\Gamma(H;U))$. The calculation of $N_*(B\Gamma(H;U))$ is achieved by the following result.

3.1.1 THEOREM. *If X is a CW complex then $N_*(X)$ is a free graded N_* module isomorphic to $H_*(X;\mathbf{Z}/2) \otimes N_*$.*

For a proof of this theorem and the subsequent one we refer the reader to section 8 of the book [Conner and Floyd, 1].

A more convenient form of this theorem will be given shortly. Let

$$\mu : N_*(X) \to H_*(X;\mathbf{Z}/2)$$

be the canonical map given by

$$\mu(M,f) = f_*\sigma$$

where σ is the fundamental homology class of M.

3.1.2 THEOREM. *Let X be a CW complex. Let singular manifolds $\{(M_a, f_a); a \in A\}$ in X be given such that $\{\mu(M_a, f_a); a \in A\}$ is a base (or generating set respectively) of $H_*(X;\mathbf{Z}/2)$. Then $\{(M_a, f_a); a \in A\}$ is a base (or generating set respectively) of $N_*(X)$ as an N_* module.*

Thus, in order to calculate a base or generating set of $N_*(B\Gamma(H;U))$ as an N_* module we shall need a base or generating set of $H_*(B\Gamma(H;U);\mathbb{Z}/2)$ or equivalently of $H^*(B\Gamma(H;U);\mathbb{Z}/2)$. This we do in the next few sections.

3.2. SIMPLIFICATION OF $\Gamma(H;U)$

We recall the definition of $\Gamma(H;U)$. Let H be a subgroup of the finite abelian group G and let U be an H module. The H module U determines a homomorphism

$$r: H \to O_k$$

where $k = \dim U$. We denote by $O(H;U)$ the centraliser of $r(H)$ in O_k. The group $\Gamma(H;U)$ is by definition

$$\Gamma(H;U) = (G \times O(H;U))/\Delta(H)$$

where $\Delta(H)$ is defined by

$$\Delta(H) = \{(h, r(h)); h \in H\}.$$

For simplicity we shall write the group $\Gamma(H;U)$ as

$$\Gamma(H;U) = (G \times O(H;U))/H$$

where we think of $G \times O(H;U)$ as a right H space with action given by

$$(g, \alpha)h = (gh, \alpha r(h)), \quad g \in G, \ \alpha \in O(H;U).$$

The group $O(H;U)$ is a product of orthogonal groups and unitary groups (see 1.6.4). Thus

$$\Gamma(H;U) = (G \times \prod_{i=1}^{m} O_{k(i)} \times \prod_{j=1}^{n} U_{\ell(j)})/H.$$

The action of H on the $O_{k(i)}$ and $U_{\ell(j)}$ may be described as follows:

(i) Given $O_{k(i)}$, there is a subgroup K_i of H of index 2 such that

$$\alpha \cdot h = \begin{cases} \alpha & \text{if } h \in K_i \\ -\alpha & \text{if } h \in H-K_i \end{cases}$$

for $\alpha \in O_{k(i)}$, $h \in H$.

(ii) Given $U_{\ell(j)}$, there is a subgroup L_j of H of index $q \neq 2$ such that H/L_j is a cyclic group. We also have a decomposition of H/L_j

$$L_j \cup xL_j \cup \ldots \cup x^{q-1}L_j$$

and a primitive q-th root of unity ξ_j such that

$$\alpha \cdot h = ((\xi_j)^k_I)\alpha \quad \text{if } h \in x^k L_j .$$

3.2.1 LEMMA.

$$\Gamma(H;U) \cong ((G \times \Pi O_{k(i)})/H) \times \Pi U_{\ell(j)} .$$

PROOF. We shall show that the action of H on each of the $U_{\ell(j)}$ extends to an action of G. Suppose we are given U_ℓ, then the action of H on U_ℓ is determined by a homomorphism

$$H \to S^1$$

(see 1.6). We claim that there is a homomorphism

$$G \to S^1$$

which extends the first.

A simple proof of this fact will be given. First of all let

us show that if ζ_n is a fixed n-th root of unity then:

> Given an element $x \in G$ of order n there exists a homomorphism $f: G \to S^1$ such that $f(x) = \zeta_n$.

If G is cyclic with generator g then we write $x = g^k$, choose $\xi = \zeta_n^{1/k}$ and define f by $f(g^j) = \xi^j$. If G has the form

$$\mathbb{Z}/n_1 \times \mathbb{Z}/n_2 \times \ldots \times \mathbb{Z}/n_r$$

then we may write x as

$$x = \prod_i x_i, \quad x_i \in \mathbb{Z}/n_i .$$

We have

$$n = \text{order}(x) = l.c.m. (\text{order}(x_1), \ldots, \text{order}(x_r))$$

and

$$\zeta_n = \xi_1 \xi_2 \ldots \xi_r$$

where ξ_i is a root of unity of order $(\text{order}(x_i))$. This reduces to the case of G cyclic and proves our assertion.

We can now prove that if G is an abelian group and H is a subgroup then:

> Given a homomorphism $\phi: H \to S^1$ there exists a homomorphism $\Phi: G \to S^1$ such that $\Phi | H = \phi$.

The proof is by induction on $|G|$. Let $x_0 \in H - \{1\}$ and $\zeta = \phi(x_0)$. We know that there exists a homomorphism $f: G \to S^1$ with $f(x_0) = \zeta$. Let K be defined by

$$K = \{x \in H;\ \phi(x) = f(x)\} .$$

Define $\phi': H/K \to S^1$ by

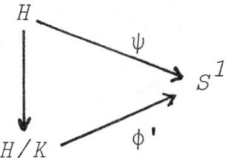

where $\psi(x) = f(x)^{-1}\phi(x)$. ($K = \ker(\psi)$.) Since $x_0 \in K$ we have $|G/K| < |G|$, by the inductive hypothesis there exists $\Phi': G/K \to S^1$ with $\Phi'|(H/K) = \phi'$. The homomorphism $\Phi: G \to S^1$ defined by

$$\Phi(x) = f(x)\Phi'(x)$$

satisfies the required condition and proves our assertion.

An alternative way of seeing that a homomorphism

$$H \to S^1$$

extends to a homomorphism

$$G \to S^1$$

is to use Lemma 57.5 in [Curtis and Reiner; p. 387] which states

"A left $\mathbb{Z}$ module M is divisible if and only if it is injective."

To explain the terminology we shall define the terms "divisible" and "injective".

"A left $\mathbb{Z}$ module M is *divisible* if $nM = M$ for all integers $n \neq 0$."

"A left $\mathbb{Z}$ module M is said to be *injective* if whenever we we are given left $\mathbb{Z}$ modules P and Q with $P \subseteq Q$, and a homomorphism η of P into M there always exists a homomorphism η' of Q into M such that η' is an extension of η, i.e., $\eta'|P = \eta$."

In our situation, H, G and S^1 are left $\mathbb{Z}$ modules and S^1 is clearly "divisible" thus any homomorphism from H to S^1 extends to a homomorphism from G to S^1.

In either way we conclude that the action of H on $\Pi U_{\ell(j)}$ extends to an action of G. An isomorphism

$$(G \times \Pi O_{k(i)} \times \Pi U_{\ell(j)})/H \to ((G \times \Pi O_{k(i)})/H) \times \Pi U_{\ell(j)}$$

is now easily obtained by using

$$(g, \beta_i, \alpha_j) \to (g, \beta_i) \times \alpha_j g^{-1}$$

for $g \in G$, $\beta_i \in O_{k(i)}$ and $\alpha_j \in U_{\ell(j)}$.

The next result simplifies groups of the form

$$(G \times \Pi O_{k(i)})/H .$$

3.2.2 LEMMA. *Let $G = \Theta \times T$ be the decomposition of G into the product of an odd order group Θ with a 2 group T. Let $H = P \times S$ be a similiar decomposition of $H \subseteq G$. Then*

$$(G \times \Pi O_{k(i)})/H \cong (\Theta/P) \times ((T \times \Pi O_{k(i)})/S) .$$

PROOF. For each O_k there is a subgroup K of H of index 2 such that H acts on O_k as

$$\alpha h = \begin{cases} \alpha & \text{if } h \in K \\ -\alpha & \text{if } h \in K-H \end{cases}$$

for $\alpha \in O_k$, $h \in H$. Because $H = P \times S$ with S a 2 group and K is a subgroup of H of index 2 we have $K = P \times S'$ where $S' \subseteq S$. Thus

$$\alpha(p,s) = \begin{cases} \alpha & \text{if } s \in S' \\ -\alpha & \text{if } s \in S-S' \end{cases}$$

for $\alpha \in O_k$, $(p,s) \in P \times S = H$.

The function

$$(\Theta \times T \times \Pi O_{k(i)})/(P \times S) \to (\Theta/P) \times ((T \times \Pi O_{k(i)})/S)$$

defined by

$$(q,t,\alpha_i) \to q \times (t,\alpha_i)$$

is well defined because of the way in which H acts on the $O_{k(i)}$. It is easily checked that this function is an isomorphism.

The next result simplies groups of the form

$$(G \times \Pi O_{k(i)})/H$$

where G and H are 2 groups.

3.2.3 LEMMA. *If G is a 2 group and H is a subgroup then*

$$(G \times \Pi O_{k(i)})/H \cong ((G/H^2) \times \Pi O_{k(i)})/(H/H^2)$$

where $H^2 = \{h^2; h \in H\}$.

PROOF. Recall that the elements of H act via multiplication by $\pm I$ on $O_{k(i)}$ and so an action of H/H^2 is well defined. (H^2 acts trivially.)

We have a homomorphism

$$G \times \Pi O_{k(i)} \to ((G/H^2) \times \Pi O_{k(i)})/(H/H^2)$$

defined by

$$(g,\alpha_i) \to (gH^2, \alpha_i)$$

for $g \in G$, $\alpha_i \in O_{k(i)}$. It is clear that it is surjective and that the kernel is precisely

$$\{(h, r(h)); h \in H\}.$$

3.2.4 REMARK. H/H^2 is an *elementary abelian 2 group* - i.e., isomorphic to a group of the form $(\mathbb{Z}/2)^k$ for some k.

3.2.5 LEMMA. *Let G be an abelian 2 group and let H be a subgroup of G so that H is an elementary abelian 2 group. There is a base $x_1, x_2, \ldots, x_n$ of G and a base $y_1, y_2, \ldots, y_m$ of H such that*

$$y_i \in \langle x_i \rangle \qquad i = 1, 2, \ldots, m.$$

PROOF. The proof is by induction on the order $|H|$ of H. We shall use additive notation for the groups.

(i) Suppose $|H| = 2$, then $H = \langle y \rangle$. Let G be

$$G = \langle x_1 \rangle \oplus \langle x_2 \rangle \oplus \ldots \oplus \langle x_n \rangle$$

with order $(x_i) = 2^{\ell(i)}$ and $0 < \ell(1) \leq \ell(2) \leq \ldots \leq \ell(n)$. Consider $\Omega_1(G)$ where

$$\Omega_1(G) = \langle 2^{\ell(1)-1} x_1 \rangle \oplus \ldots \oplus \langle 2^{\ell(n)-1} x_n \rangle.$$

Then $\Omega_1(G)$ contains all elements of order 2, thus

$$y = \sum_{i=1}^{n} \lambda_i \, 2^{\ell(i)-1} x_i$$

with $\lambda_i = 0$ or 1. We may assume that $\lambda_i = 1$ by simply ignoring those x_i for which $\lambda_i = 0$. Thus

$$y = \sum_{i=1}^{n} 2^{\ell(i)-1} x_i .$$

Let $x_1' = \sum_{i=1}^{n} 2^{\ell(i)-\ell(1)} x_i$, then the order of x_1' is $2^{\ell(1)}$ and $y = 2^{\ell(1)-1} x_1'$. Since

$$x_1' = x_1 + \sum_{i=2}^{n} 2^{\ell(i)-\ell(1)} x_i$$

we have

$$G = <x_1'> \oplus <x_2> \oplus \ldots \oplus <x_n>$$

and

$$y = 2^{\ell(1)-1} x_1' \in <x_1'> .$$

(ii) If $|H| = 2^m > 2$ then write $H = <y> \oplus J$. By induction there is a base $x_1, x_2, \ldots, x_n$ of G and a base $y_1, y_2, \ldots, y_{m-1}$ of J such that

$$y_i \in <x_i> \qquad i = 1, 2, \ldots, m-1.$$

Now, as before,

$$y = \sum_{i=1}^{n} \lambda_i \, 2^{\ell(i)-1} x_i .$$

But $H = <y'> \oplus J$ where

$$y' = y - \sum_{i=1}^{m-1} \lambda_i \, 2^{\ell(i)-1} x_i$$

$$= \sum_{i=m}^{n} \lambda_i \, 2^{\ell(i)-1} x_i$$

and

$$\langle y' \rangle \subseteq \langle x_m \rangle \oplus \langle x_{m+1} \rangle \oplus \ldots \oplus \langle x_n \rangle$$

which after using case (i) again proves the result.

3.2.6 COROLLARY. *The group* $\Gamma(H;U)$ *is of the form*

$$\Theta \times \Pi U_{\ell(j)} \times (\langle x_1 \rangle \oplus \ldots \oplus \langle x_n \rangle \times \Pi O_{k(i)}) / (\langle y_1 \rangle \oplus \ldots \oplus \langle y_m \rangle)$$

where Θ *is a group of odd order, the* $\langle x_i \rangle$ $i = 1, \ldots, n$ *are cyclic groups of order a power of 2, the* $\langle y_i \rangle$, $i = 1, 2, \ldots, m$ *are cyclic groups of order 2 with* $\langle y_i \rangle \subseteq \langle x_i \rangle$. *The action of* y_j *on* $O_{k(i)}$ *is either trivial or multiplication by* -1.

In fact the last term could be written as

$$(\langle x_1 \rangle \oplus \ldots \oplus \langle x_n \rangle \times \prod_{i=0}^{2^m - 1} O_{k(i)}) / (\langle y_1 \rangle \oplus \ldots \oplus \langle y_m \rangle)$$

with k_i possibly zero (and hence $O_{k(i)} = pt$) and y_i acts on $O_{k(j)}$ as $+1$ or -1 respectively as the i-th entry in the 2 adic expansion of j is 0 or 1 respectively.

It is more convenient to write the above term as

$$\langle x_1 \rangle \oplus \ldots \oplus \langle x_n \rangle \times_{\langle y_1 \rangle \oplus \ldots \oplus \langle y_n \rangle} \prod_j O_{k(j)}$$

or

$$\prod_{i=1}^{n} \mathbb{Z}/2^{r(i)} \times_{(\mathbb{Z}/2)^m} \Pi O_{k(j)} \quad .$$

There is one further simplification that we can do and this involves the groups O_k with k odd. We state the result in one case, leaving the general case for the reader.

79

3.2.7 LEMMA.

$$\mathbb{Z}/2^r \times_{\mathbb{Z}/2} O_{2n+1} \cong \mathbb{Z}/2^r \times SO_{2n+1}.$$

PROOF. The isomorphism is obtained by using the following

$$(g, \alpha) \to g(\det \alpha) \times ((\det \alpha)I)\alpha$$

for $g \in \mathbb{Z}/2^r$, $\alpha \in O_{2n+1}$, where $\det \alpha$ denotes the determinant of the matrix α.

REMARK. $O_{2n+1} \cong \mathbb{Z}/2 \times SO_{2n+1}$.

3.3 COHOMOLOGY OF $B\Gamma$

From the last section and the fact that

$$H^*(B(\Gamma_1 \times \Gamma_2), \mathbb{Z}/2) \cong H^*(B\Gamma_1; \mathbb{Z}/2) \otimes H^*(B\Gamma_2; \mathbb{Z}/2)$$

we see that the cohomology of the classifying spaces of the following groups are needed.

(i) Γ = an odd order abelian group.

(ii) a) $\Gamma = \mathbb{Z}/2$, b) $\Gamma = \mathbb{Z}/2^r$, $r > 1$.

(iii) a) $\Gamma = O_k$. b) $\Gamma = SO_k$.

(iv) $\Gamma = U_\ell$.

(v) $\Gamma = \mathbb{Z}/4 \times_{\mathbb{Z}/2} O_k$.

(vi) $\Gamma = \mathbb{Z}/2^r \times_{\mathbb{Z}/2} O_k$.

(vii) $\Gamma = \prod_{i=1}^{n} \mathbb{Z}/2^{r_i} \times (\mathbb{Z}/2)^m \prod_{j=1}^{2^m-1} O_{k(j)}$.

The cohomology of the classifying spaces of the first four types of group is well known and appears in many books on algebraic topology. We shall merely state the result. The results for cases (v) and (vi) (and (ii)b) will be obtained in this section. The case (vii) will not be attempted (for $m > 1$).

3.3.1 THEOREM. *(i) If Γ is a finite abelian group of odd order then*

$$H^*(B\Gamma; \mathbb{Z}/2) \cong \mathbb{Z}/2$$

(ii) If $\Gamma = \mathbb{Z}/2$ then

$$H^*(B\mathbb{Z}/2; \mathbb{Z}/2) \cong \mathbb{Z}/2[\alpha]$$

where α is the universal Stiefel-Whitney class of the associated line bundle over $B\mathbb{Z}/2$.

(iii) (a) $\Gamma = O_k$.

$$H^*(BO_k; \mathbb{Z}/2) \cong \mathbb{Z}/2[w_1, w_2, \ldots, w_k]$$

where the w_i are the universal Stiefel-Whitney classes.

(iii) (b) $\Gamma = SO_k$.

$$H^*(BSO_k; \mathbb{Z}/2) \cong \mathbb{Z}/2[w_2, w_3, \ldots, w_k].$$

(iv) $\Gamma = U_\ell$.

$$H^*(BU_\ell; \mathbb{Z}/2) \cong \mathbb{Z}/2[c_1, c_2, \ldots, c_\ell]$$

where the c_i are the universal Chern classes.

3.3.2 THEOREM. *The cohomology of $B\mathbb{Z}/2^r$, $r > 1$ is*

$$\mathbb{Z}/2[\alpha, \beta]/\langle \alpha^2 \rangle .$$

The classes α, β have the following description. Consider the maps in cohomology induced by the fibration

$$B\mathbb{Z}/2 \to B\mathbb{Z}/2^r \to B\mathbb{Z}/2^{r-1} .$$

The class α comes from α in $H^*(B\mathbb{Z}/2^{r-1}; \mathbb{Z}/2)$

$$H^*(B\mathbb{Z}/2^{r-1};\mathbb{Z}/2) = \begin{cases} \mathbb{Z}/2\,[\alpha,\beta]/<\alpha^2> & \text{if } r > 2 \\ \mathbb{Z}/2\,[\alpha] & \text{if } r = 2 \end{cases}$$

while the class β restricts to α^2 in $\mathbb{Z}/2\,[\alpha] = H^*(B\mathbb{Z}/2;\mathbb{Z}/2)$.

A proof of the above theorem will appear naturally in the calculation of the cohomology of $B\Gamma$ when

$$\Gamma = \mathbb{Z}/2^r \times_{\mathbb{Z}/2} O_k$$

which for $k = 1$ is $\mathbb{Z}/2^r$ (and for $k = 0$ is $\mathbb{Z}/2^{r-1}$).

To state the result concerning the cohomology of $B\Gamma$, where

$$\Gamma = \mathbb{Z}/2^r \times_{\mathbb{Z}/2} O_k$$

we need only consider the case k even since for k odd the result follows from 3.2.7, 3.3.1 and 3.3.2. For k even we shall introduce some classes in $H^*(B\Gamma)$.

3.3.3 Let $\Gamma = \mathbb{Z}/2^r \times_{\mathbb{Z}/2} O_{2n}$ and consider the fibration

$$BO_{2n} \to B\Gamma \to B\mathbb{Z}/2^{r-1}\;.$$

There are the following classes in $H^*(B\Gamma)$.

(i) $\alpha \in H^1$ which comes from $\alpha \in H^1(B\mathbb{Z}/2^{r-1})$. If $r > 2$ then $\alpha^2 = 0$.

(ii) $\beta \in H^2$ which comes from $\beta \in H^2(B\mathbb{Z}/2^{r-1})$. If $r = 2$ $\alpha^2 = \beta$.

(iii) $u_{2j-1} \in H^{2j-1}$ which restricts to $w_{2j-1} \in H^{2j-1}(BO_{2n})$, $j = 1,2,\ldots,n$.

(iv) $v_{4j} \in H^{4j}$ which restricts to w_{2j}^2 in $H^{4j}(BO_{2n})$, $j = 1,2,\ldots,n$.

(v) $\rho_S \in H^{|S|-1}$ which restricts to a class

$\rho_S \in H^{|S|-1}(BO_{2n})$, $S \in S(2,4,6,\ldots,2n)$, $S \neq \emptyset$, where $S(2,4,\ldots,2n)$ denotes the set of subsets of $\{2,4,\ldots,2n\}$ excluding subsets consisting of a single element. If $S \in S(2,4,\ldots,2n)$ then $S = (s(1),s(2),\ldots,s(k))$ for some $k \neq 1$ and

$$2 \leq s(1) < s(2) < \ldots < s(k) \leq 2n$$

with each $s(i)$ even, the class ρ_S is defined as

$$\rho_S = \sum_{i=1}^{k} w_{s(1)} w_{s(2)} \cdots w_{s(i-1)} w_{s(i)-1} w_{s(i+1)} \cdots w_{s(k)}.$$

3.3.4 DEFINITION. Let $\sigma(i,j,K,L,S)$ be the class in $H^*(B(\mathbb{Z}/2^r \times_{\mathbb{Z}/2} O_{2n}))$ given by

$$\alpha^i \beta^j u_1^{k(1)} u_3^{k(2)} \cdots u_{2n-1}^{k(n)} v_4^{\ell(1)} v_8^{\ell(2)} \cdots v_{4n}^{\ell(n)} \rho_S$$

where $K = (k(1),k(2),\ldots,k(n))$ and $L = (\ell(1),\ell(2),\ldots,\ell(n))$ are n-tuples of non-negative integers, $S \in S(2,4,6,\ldots,2n)$ and if $S = \emptyset$ then $\rho_S = 1$.

3.3.5 THEOREM. Let $\Gamma = \mathbb{Z}/2^r \times_{\mathbb{Z}/2} O_{2n}$, then a $\mathbb{Z}/2$ basis for $H^*(B\Gamma; \mathbb{Z}/2)$ is given by

$$\begin{cases} \sigma(0,0,K,L,S) \\ \sigma(1,0,K,L,S) & \text{if } r = 2 \\ \sigma(i,0,0,L,0) \ i > 1 \end{cases}$$

$$\begin{cases} \sigma(0,0,K,L,S) \\ \sigma(1,0,K,L,S) \\ \sigma(0,j,0,L,0) \quad j \geq 1 \\ \sigma(1,j,0,L,0) \quad j \geq 1 \end{cases} \quad if \; r > 2$$

where i,j are integers, K,L are n-tuples of non-negative integers and $S \in S(2,4,\ldots,2n)$ the set of subsets of $\{2,4,\ldots,2n\}$ except those consisting of a single element.

We shall devote the rest of this section to the calculation of the cohomology of $B(\mathbb{Z}/2^r \times_{\mathbb{Z}/2} O_k)$. This will be achieved by using the Serre spectral sequence. Recall that the Serre spectral sequence of a fibration

$$F \to E \to B$$

takes the form

$$E_2^{*,*} = H^*(B; H^*(F, \mathbb{Z}/2)) \cong H^*(B; \mathbb{Z}/2) \otimes H^*(F; \mathbb{Z}/2)$$
$$\Rightarrow H^*(E; \mathbb{Z}/2)$$

if the fibration is "orientable with respect to $H^*(-;\mathbb{Z}/2)$", (see, for example, [Spanier; p.476]). We claim that the fibration

$$BO_k \to B(\mathbb{Z}/2^r \times_{\mathbb{Z}/2} O_k) \to B\mathbb{Z}/2^{r-1}$$

is orientable with respect to H^*. To prove this, first consider the fibration

$$BO_k \to B(S^1 \times_{\mathbb{Z}/2} O_k) \to BS^1/(\mathbb{Z}/2)$$

since this has a simply connected base space it is orientable with respect to H^* (see [Spanier; p.476]).

There is a map
$$B((\mathbb{Z}/2^r)/(\mathbb{Z}/2)) \xrightarrow{f} B(S^1/(\mathbb{Z}/2))$$

since $\mathbb{Z}/2^r \subseteq S^1$ and the induced fibration is precisely

$$BO_k \to B(\mathbb{Z}/2^r \times_{\mathbb{Z}/2} O_k) \to B\mathbb{Z}/2^{r-1}.$$

(This fibration maps into the induced fibration, but the maps on the base and fibre are the identity maps.) Finally we use the fact [Spanier; p.476] that a fibration induced from a fibration orientable with respect to H^* is itself orientable with respect to H^*.

We start by looking at the case $r = 2$. Consider the following fibrations

$$BO_k \to B(\mathbb{Z}/4 \times_{\mathbb{Z}/2} O_k) \xrightarrow{\pi} B\mathbb{Z}/2$$

and

$$B(\mathbb{Z}/2)^k \to B(\mathbb{Z}/4 \times_{\mathbb{Z}/2} (\mathbb{Z}/2)^k) \xrightarrow{\pi^k} B\mathbb{Z}/2 .$$

We have already shown that the first fibration is orientable with respect to H^*, that the second one is also orientable with respect to H^* is easy to see.

We have a commutative diagram of fibrations

$$\begin{array}{ccccc} B(\mathbb{Z}/2)^k & \to & B(\mathbb{Z}/4 \times_{\mathbb{Z}/2} (\mathbb{Z}/2)^k) & \xrightarrow{\pi^k} & B\mathbb{Z}/2 \\ \downarrow f & & \downarrow g & & \downarrow 1 \\ BO_k & \to & B(\mathbb{Z}/4 \times_{\mathbb{Z}/2} O_k) & \xrightarrow{\pi} & B\mathbb{Z}/2 \end{array}$$

with f being induced by the inclusion $(\mathbb{Z}/2)^k \subset O_k$ as the maximal torus.

The map $f^*: H^*(BO_k) \to H^*(B(\mathbb{Z}/2)^k)$ is injective and is given by

$$H^*(BO_k) \longrightarrow H^*(B(\mathbb{Z}/2)^k)$$

$$\mathbb{Z}/2[w_1,\ldots,w_k] \longrightarrow \mathbb{Z}/2[t_1, t_2,\ldots,t_k]$$

$$w_i \longmapsto \sigma_i$$

where σ_i is the i-th elementary symmetric function of $t_1, t_2,\ldots,t_k$. We conclude that

$$g^*: E_2^{**}(\pi) \to E_2^{**}(\pi^k)$$

is a monomorphism because

$$E_2^{**}(\pi) = H^*(B\mathbb{Z}/2) \otimes H^*(BO_k)$$
$$\cong \mathbb{Z}/2[\alpha] \otimes \mathbb{Z}/2[w_1,\ldots,w_k]$$

and

$$E_2^{**}(\pi^k) = H^*(B\mathbb{Z}/2) \otimes H^*(B(\mathbb{Z}/2)^k)$$
$$= \mathbb{Z}/2[\alpha] \otimes \mathbb{Z}/2[t_1, t_2,\ldots,t_k] \quad .$$

Consider now the spectral sequence of π^k, i.e., $E_2^{**}(\pi^k)$. We have, of course, $d_2(\alpha) = 0$. Now $d_2(t_i)$ is α^2 or 0.

3.3.6 LEMMA. $d_2(t_i) = \alpha^2$, $i = 1, 2,\ldots,k$.

PROOF. Suppose that $d_2(t_1) = 0$, then $t_1 = i^*x$ for some $x \in H^1(B(\mathbb{Z}/4 \times_{\mathbb{Z}/2} (\mathbb{Z}/2)^k))$ where i is the natural inclusion

$$i: B(\mathbb{Z}/2^k) \to B(\mathbb{Z}/4 \times_{\mathbb{Z}/2} (\mathbb{Z}/2)^k) \quad .$$

(This follows from the exactness of the sequence

$$H^1(E) \xleftarrow{i^*} H^1(F) \xrightarrow{d_2} H^2(B)$$

in Serre spectral sequence, see for example, [Hilton and Wylie; p.431].)

Consider the following principal $(\mathbb{Z}/2)^k$ bundle

$$S^1 \times (\mathbb{Z}/2)^{k-1}$$
$$\downarrow$$
$$RP^1$$

where the first factor of $(\mathbb{Z}/2)^k$ acts antipodally on S^1, all other factors act trivially. The action on $(\mathbb{Z}/2)^{k-1}$ being the obvious one. Let

$$h: RP^1 \to B(\mathbb{Z}/2)^k$$

classify this principal $(\mathbb{Z}/2)^k$ bundle. There is a principal $\mathbb{Z}/4 \times_{\mathbb{Z}/2} (\mathbb{Z}/2)^k$ bundle

$$S^1 \times (\mathbb{Z}/2)^{k-1}$$
$$\downarrow$$
$$RP^1/(\mathbb{Z}/2)$$

where the generator of $\mathbb{Z}/4$ acts on S^1 by multiplication by $i = \sqrt{-1}$ and on each of the factors of $(\mathbb{Z}/2)^{k-1}$ by multiplication by -1. The action of $(\mathbb{Z}/2)^k$ on $S^1 \times (\mathbb{Z}/2)^{k-1}$ is as described previously. Let

$$g: RP^1/(\mathbb{Z}/2) \to B(\mathbb{Z}/4 \times_{\mathbb{Z}/2} (\mathbb{Z}/2)^k)$$

classify this principal bundle. If f

$$f: RP^1 \to RP^1/(\mathbb{Z}/2)$$

denotes the canonical projection then we have a commutative diagram

$$\begin{array}{ccc} RP^1 & \xrightarrow{h} & B(\mathbb{Z}/2)^k \\ \downarrow f & & \downarrow i \\ RP^1/(\mathbb{Z}/2) & \xrightarrow{g} & B(\mathbb{Z}/4 \times_{\mathbb{Z}/2} (\mathbb{Z}/2)^k) \end{array}.$$

The assumption $i^*x = t_1$ leads to

$$h^*i^*x = h^*t_1 = w_1(\xi)$$
$$\|$$
$$f^*g^*x$$

where $w_1(\xi)$ denotes the first Stiefel-Whitney class of the canonical line bundle over RP^1. Thus we have

$$<f^*g^*x, RP^1> = <w_1(\xi), RP^1> = 1$$
$$\|$$
$$<g^*x, f_*RP^1>$$
$$\|$$
$$<g^*x, 0>$$
$$\|$$
$$0$$

since f is a double covering. This leads to a contradiction and hence $d_2(t_1) = \alpha^2$. Similiarly we deduce that $d_2(t_i) = \alpha^2$ for $i = 1, 2, \ldots, k$.

To continue we shall need to know what happens to the elementary symmetric functions σ_i under the differential $d = d_2$.

3.3.7 LEMMA.
$$d(\sigma_i(t_1, t_2, \ldots, t_k)) = \begin{cases} 0 & \text{if } i+k \text{ is odd} \\ \alpha^2 \sigma_{i-1}(t_1, \ldots, t_k) & \text{if } i+k \text{ is even and } i > 0 \end{cases}$$

PROOF. The proof is by induction on k. If $k = 1$ then we have

$$d(\sigma_0(t_1)) = 0$$

$$d(\sigma_1(t_1)) = d(t_1) = \alpha^2 = \alpha^2 \sigma_0(t_1),$$

which verifies the result.

Suppose now that $k > 1$. We have

$$\sigma_i(t_1, t_2, \ldots, t_k) = \sigma_i(t_1, \ldots, t_{k-1}) + t_k \sigma_{i-1}(t_1, \ldots, t_{k-1})$$

and so

$$d(\sigma_i(t_1, \ldots, t_k)) = d(\sigma_i(t_1, \ldots, t_{k-1}))$$
$$+ t_k \, d(\sigma_{i-1}(t_1, \ldots, t_{k-1}))$$
$$+ \alpha^2 \sigma_{i-1}(t_1, \ldots, t_{k-1}).$$

If $(i+k)$ is odd then $i+(k-1)$ is even and $(i-1)+(k-1)$ is odd, so we obtain

$$d(\sigma_i(t_1, \ldots, t_k)) = \alpha^2 \sigma_{i-1}(t_1, \ldots, t_{k-1}) + 0$$
$$+ \alpha^2 \sigma_{i-1}(t_1, \ldots, t_{k-1})$$
$$= 0.$$

Whereas if $i+k$ is even then $i+(k-1)$ is odd and $(i-1)+(k-1)$ is even so that

$$d(\sigma_i(t_1, \ldots, t_k)) = 0 + t_k \alpha^2 \sigma_{i-2}(t_1, \ldots, t_{k-1})$$
$$+ \alpha^2 \sigma_{i-1}(t_1, \ldots, t_{k-1})$$
$$= \alpha^2 \sigma_{i-1}(t_1, \ldots, t_k)$$

which establishes the induction.

This result together with the injectivity of $E_2^{*,*}(\pi) \to E_2^{*,*}(\pi^k)$ establishes the next result.

3.3.8 COROLLARY. *In* $E_2^{*,*}(\pi)$ *we have* $d(\alpha) = 0$ *and*

$$d(w_i) = \begin{cases} 0 & \text{if } i+k \text{ is odd} \\ \alpha^2 w_{i-1} & \text{if } i+k \text{ is even.} \end{cases}$$

NOTE. If k is odd then it is very easy to see that

$$E_3^{**}(\pi) \cong \mathbb{Z}/2\left[\alpha, w_1^2, w_2, w_3+w_1 w_2, w_4, \ldots, w_{k-1}, w_k+w_1 w_{k-1}\right]/\langle \alpha^2 \rangle,$$

and of course this is also the E_∞ term.

To calculate the E_3^{**} term in the case that k is even first look at the next lemma.

3.3.9 LEMMA. *Suppose* $k = 2n$ *and suppose that* $f(\alpha, w_1, w_2, \ldots, w_{2n})$ *is an element of* $E_2^{**} = \mathbb{Z}/2\left[\alpha, w_1, w_2, \ldots, w_{2n}\right]$. *If* $df = 0$ *then*

$$\alpha^2 f(\alpha, w_1, \ldots, w_k) = \alpha^2 g(\alpha, w_2^2, w_4^2, \ldots, w_{2n}^2) + d(h(\alpha, w_1, \ldots, w_{2n}))$$

for some polynomials g, h.

PROOF. We use induction on n. If $k = 2n = 2$ then

$$f(\alpha, w_1, w_2) = \sum_{a,b \geq 0} w_1^a w_2^b P_{a,b}$$

where $P_{a,b}$ is a polynomial in α. Next we have

$$df = \sum_{\substack{a \geq 0 \\ b \geq 1}} b\alpha^2 w_1^{a+1} w_2^{b-1} P_{a,b} \ .$$

Now $df = 0$ if and only if b is even always, i.e.,

$$f = \sum_{a,c} w_1^a w_2^{2c} P_{a,2c}$$

$$= \sum_{c \geq 0} w_2^{2c} P_{0,2c} + \sum_{c \geq 0, a \geq 0} w_1^a w_2^{2c} P_{a,2c}$$

But,
$$d(w_1^a w_2^{2c+1}) = \alpha^2 w_1^a w_2^{2c}$$

and so
$$d(w_1^a w_2^{2c+1} P_{a,2c}) = \alpha^2 w_1^a w_2^{2c} P_{a,2c}.$$

Hence,
$$\alpha^2 f = \sum \alpha^2 w_2^{2c} P_{0,2c} + d(\sum w_1^a w_2^{2c+1} P_{a,2c})$$

which establishes the result for $k = 2n = 2$.

Suppose now that $k = 2n \geq 4$. We can write f as
$$f(\alpha, w_1, w_2, \ldots, w_{2n}) = \sum_{a,c \geq 0} w_1^a w_2^{2c} P_{a,2c} + \sum_{a,c \geq 0} w_1^a w_2^{2c+1} P_{a,2c+1}$$

where $P_{a,2c}, P_{a,2c+1}$ are polynomials in the variables $\alpha, w_3, w_4, \ldots, w_{2n}$. Now $df = 0$ means that
$$\sum w_1^a w_2^{2c} dP_{a,2c} + \sum \alpha^2 w_1^{a+1} w_2^{2c} P_{a,2c+1} + \sum w_1^a w_2^{2c+1} dP_{a,2c+1} = 0$$

or, in other words
$$\alpha^2 P_{a,2c+1} = dP_{a+1,2c} \quad \text{for } a,c \geq 0$$

(and $dP_{a,2c+1} = 0$). Thus we may write $\alpha^2 f$ as
$$\alpha^2 \sum w_1^a w_2^{2c} P_{a,2c} + \sum w_1^a w_2^{2c+1} dP_{a+1,2c}.$$

But
$$\sum_{c \geq 0} \alpha^2 w_2^{2c} P_{0,2c} + \sum_{\substack{a \geq 1 \\ c \geq 0}} d(w_1^{a-1} w_2^{2c+1} P_{a,2c})$$
$$= \sum_{a,c \geq 0} \alpha^2 w_1^a w_2^{2c} P_{a,2c} + \sum_{\substack{a \geq 1 \\ c \geq 0}} w_1^{a-1} w_2^{2c+1} dP_{a,2c}$$
$$= \alpha^2 f.$$

Since $df = 0$ we see that $dP_{0,2c} = 0$ and by induction

$$\alpha^2 P_{0,2c} = \alpha^2 p(\alpha, w_4^2, w_6^2, \ldots, w_{2n}^2) + d(q(\alpha, w_3, w_4, \ldots, w_{2n})).$$

So

$$\alpha^2 f = \alpha^2 \sum w_c^{2c} p(\alpha, w_4^2, w_6^2, \ldots, w_{2n}^2)$$

$$+ \alpha^2 \sum w_c^{2c} d(q(\alpha, w_3, \ldots, w_{2n}))$$

$$+ \sum d(w_1^{a-1} w_2^{2c+1} P_{a,2c})$$

which is of the required form

$$\alpha^2 g(\alpha, w_2^2, w_4^2, \ldots, w_{2n}^2) + d(h(\alpha, w_1, w_2, \ldots, w_{2n}))$$

since $\alpha^2 \sum w_c^{2c} dq = d(\sum \alpha^2 w_2^{2c} q)$.

This lemma immediately gives us the next result.

3.3.10 COROLLARY. *If $k = 2n$ then*

$$\bigoplus_{\substack{q \geq 0 \\ p \geq 2}} E_3^{p,q}(\pi) \cong \alpha^2 \mathbb{Z}/2[\alpha, w_2^2, w_4^2, \ldots, w_{2n}^2] \quad .$$

3.3.11 LEMMA. $E_3^{**} = E_\infty^{**}$ *for $k = 2n$.*

PROOF. There is a diagram of fibrations

$$\begin{array}{ccccc}
BU_n & \longrightarrow & B(\mathbb{Z}/4 \times_{\mathbb{Z}/2} U_n) & \xrightarrow{\pi^U} & B\mathbb{Z}/2 \\
\downarrow & & \downarrow & & \downarrow 1 \\
BO_{2n} & \longrightarrow & B(\mathbb{Z}/4 \times_{\mathbb{Z}/2} O_{2n}) & \xrightarrow{\pi} & B\mathbb{Z}/2
\end{array} \quad .$$

Because $\mathbb{Z}/4 \times_{\mathbb{Z}/2} U_n \cong \mathbb{Z}/2 \times U_n$ (see Lemma 4.3.1) the spectral sequence of π^U collapses and so we have

$$E_r^{**}(\pi^U) \cong \mathbb{Z}/2[\alpha, c_1, c_2, \ldots, c_n]$$

for all r. The homomorphism

$$\Theta_2 : E_2^{**}(\pi) \to E_2^{**}(\pi^U)$$

defined by

$$\Theta_2(\alpha) = \alpha$$
$$\Theta_2(w_{2i-1}) = 0$$
$$\Theta_2(w_{2i}) = c_i$$

commutes with the differentials and hence induces homomorphisms

$$\Theta_r : E_r^{**}(\pi) \to E_2^{**}(\pi^U) \ .$$

From Corollary 3.3.10 we see that Θ_3 is a monomorphism when restricted to $\bigoplus_{\substack{q \geq 0 \\ p \geq 2}} E_3^{p,q}(\pi)$. Thus $\bigoplus_{\substack{q \geq 0 \\ p \geq 2}} E_3^{p,q}(\pi)$ is in the kernel of d_3. Now, if $x = d_3 y$ then x is in $\bigoplus_{p \geq 2} E_3^{p,q}(\pi)$ since $d_3 : E_3^{p,q} \to E_3^{p+3, q-2}$. Also, since

$$\Theta_3(x) = d_3(\Theta_3(y)) = 0$$

we have $x = 0$ and hence $d_3 = 0$. By induction on r we can show that Θ_r is monic on $\bigoplus_{p \geq 2} E_r^{p,q}(\pi)$ and that $d_r = 0$ for all $r \geq 3$ thereby proving the assertion.

We still need to calculate $E_3^{0,*}$ and $E_3^{1,*}$. The next lemma gives a solution. First we recall some definitions given earlier in 3.3.3.

3.3.12 DEFINITION. Let $S_{2n} = S(2, 4, \ldots, 2n)$ denote the set of subsets of $\{2, 4, \ldots, 2n\}$ excluding the subsets consisting of one element alone.

Thus $S \in S_{2n}$ means that $S = (s(1), s(2), \ldots, s(k))$ for some $k \neq 1$ and

$$2 \leq s(1) < s(2) < \ldots < s(k) \leq 2n$$

with each $s(i)$, $i = 1, 2, \ldots, k$, an even integer.

Given $S \in S_{2n}$ we define $\rho_S \in \mathbb{Z}/2[w_1, w_2, \ldots, w_{2n}]$ by the equation

$$\alpha^2 \rho_S = d(w_{s(1)} w_{s(2)} \cdots w_{s(k)})$$

if $S = (s(1), s(2), \ldots, s(k))$ with $k > 1$. In other words

$$\rho_S = \sum_{i=1}^{k} w_{s(1)} \cdots w_{s(i-1)} w_{s(i)-1} w_{s(i+1)} \cdots w_{s(k)} .$$

If $k = 0$, i.e., $S = \emptyset$ then we define $\rho_S = 1$.

3.3.13 LEMMA. *Suppose that* $f \in \mathbb{Z}/2[w_1, w_2, \ldots, w_{2n}]$. *If* $df = 0$ *then*

$$f = \sum \rho_S g_S(w_1, w_2^2, w_3, w_4^2, \ldots, w_{2n}^2)$$

for some polynomials g_S, *the sum being taken over* $S \in S_{2n}$.

PROOF. The proof is by induction on n. If $n = 1$ then

$$f = \sum w_1^a w_2^b P_{a,b}$$

for some $P_{a,b} \in \mathbb{Z}/2$. Thus

$$df = \sum \alpha^2 b w_1^{a+1} w_2^b P_{a,b}$$

so that $df = 0$ means that $b P_{a,b} = 0$ for all b. Hence

94

$$f = \sum w_1^a w_2^{2c} P_{a,2c}$$
$$= g_\emptyset(w_1, w_2^2) \quad .$$

(Note that $S_2 = \{\emptyset\}$.)

Now suppose that $n > 1$. We represent f as

$$f = \sum w_1^a w_2^{2c} P_{a,2c} + \sum_{a \geq 1} w_1^{a-1} w_2^{2c+1} P_{a-1, 2c+1}$$

where $P_{a,b} \in \mathbb{Z}/2[w_3, w_4, \ldots, w_{2n}]$. Since $df = 0$ we see that $dP_{a-1, 2c+1} = 0$ and by the inductive hypothesis we have

$$P_{a-1, 2c+1} = \sum_S \rho_S g_{a,c,S}(w_3, w_4^2, \ldots, w_{2n}^2)$$

where $S \in S(4, 6, \ldots, 2n)$. Thus

$$f = \sum w_1^a w_2^{2c} P_{a,2c} + \sum w_1^{a-1} w_2^{2c+1} \rho_S g_{a,c,S} \quad .$$

Let $w_S = w_{s(1)} w_{s(2)} \cdots w_{s(k)}$ if $S = (s(1), s(2), \ldots, s(k))$. The polynomial $P_{a,2c}$ may be written as

$$P_{a,2c} = \sum_S w_S h_{a,c,S}(w_3, w_4^2, \ldots, w_{2n}^2) + \sum_{i=2}^n w_{2i} h_{a,c,2i}(w_3, w_4^2, \ldots, w_{2n}^2)$$

where $S \in S(4, 6, \ldots, 2n)$ and $h_{a,c,S}, h_{a,c,2i}$ are polynomials. (This holds in fact for any polynomial in $\mathbb{Z}/2[w_3, w_4, \ldots, w_{2n}]$).

Because $df = 0$ we have

$$dP_{a,2c} = \alpha^2 P_{a-1, 2c+1}$$
$$dP_{0,2c} = 0$$

in other words

$$\sum_S d(w_S) h_{a,c,S} + \sum_{i=2}^n \alpha^2 w_{2i-1} h_{a,c,2i}$$
$$= \sum_S \alpha^2 \rho_S g_{a,c,S} \qquad (a \geq 1).$$

Since the elements ρ_S, $S \in S(4,6,\ldots,2n)$ are independent in $\mathbb{Z}/2\left[w_3, w_4^2, \ldots, w_{2n}^2\right]$ we deduce that

$$h_{a,c,S} = g_{a,c,S} \qquad a \geq 1, \ S \neq \emptyset$$

$$\sum_{i=2}^{n} w_{2i-1} h_{a,c,2i} = g_{a,c,\emptyset} \ .$$

So

$$f = \sum_{\substack{a \geq 1 \\ S \neq \emptyset}} w_1^a w_2^{2c} w_S g_{a,c,S} + \sum w_2^{2c} P_{0,2c}$$

$$+ \sum_{\substack{i \geq 2 \\ a \geq 1}} w_1^a w_2^{2c} w_{2i} h_{a,c,2i}$$

$$+ \sum w_1^a w_2^{2c} h_{a,c,\emptyset}$$

$$+ \sum_{a \geq 1} w_1^{a-1} w_2^{2c+1} \rho_S g_{a,c,S}$$

$$= \sum_{\substack{a \geq 1 \\ S \neq \emptyset}} (w_1 w_S + w_2 \rho_S) w_1^{a-1} w_2^{2c} g_{a,c,S} + \sum w_2^{2c} P_{0,2c}$$

$$+ \sum_{a \geq 1} \left(\sum_{i=2}^{n} w_1 w_{2i} h_{a,c,2i} + w_2 g_{a,c,\emptyset} \right) w_1^{a-1} w_2^{2c}$$

$$+ \sum w_1^a w_2^{2c} h_{a,c,\emptyset} \ .$$

Since $w_1 w_S + w_2 \rho_S = \rho_{S'}$, where $S' = \{2\} \cup S$ we see that the first term is of the required form. The second term is of the required form by the inductive hypothesis since $dP_{0,2c} = 0$. For the third term we see that

$$\sum_{i=2}^{n} w_1 w_{2c} h_{a,c,2i} + w_2 g_{a,c,\emptyset}$$

$$= \sum_{i=2}^{n} (w_1 w_{2c} h_{a,c,2i} + w_2 w_{2i-1} h_{a,c,2i})$$

$$= \sum_{i=2}^{n} (w_1 w_{2i} + w_2 w_{2i-1}) h_{a,c,2i}$$

$$= \sum_{i=2}^{n} \rho_{\{2,2i\}} h_{a,c,2i}$$

so that the third term is also of the required form. Finally the last term is clearly of the required form. This completes the induction and proof of the lemma.

3.3.14 COROLLARY. *(i)* $E_\infty^{0,*}$ *is the* $\mathbb{Z}[w_1, w_2^2, w_3, \ldots, w_{2n}^2]$ *module on generators* ρ_S, $S \in S_{2n}$

(ii) $E_\infty^{*,*}$ *is the* $\mathbb{Z}[\alpha, w_1, w_2^2, w_3, \ldots, w_{2n}^2]$ *module on generators* ρ_S, $S \in S_{2n}$ *and relations*

$$\alpha^2 w_{2i-1} = \alpha^2 \rho_S = 0, \quad (S \neq \emptyset).$$

From this corollary we obtain Theorem 3.3.5 for the case $r = 2$ i.e., for $\Gamma = \mathbb{Z}/4 \times_{\mathbb{Z}/2} O_{2n}$.

The calculation of the cohomology of $B\Gamma$ when Γ is $\mathbb{Z}/2^r \times_{\mathbb{Z}/2} O_k$ is performed in a similiar fashion as for $\mathbb{Z}/4 \times_{\mathbb{Z}/2} O_k$. We have a diagram of fibrations

$$\begin{array}{ccccc} BO_k & \to & B(\mathbb{Z}/4 \times_{\mathbb{Z}/2} O_k) & \xrightarrow{\pi} & B\mathbb{Z}/2 \\ \downarrow & & \downarrow \Theta & & \downarrow \\ BO_k & \to & B(\mathbb{Z}/2^r \times_{\mathbb{Z}/2} O_k) & \xrightarrow{\pi_r} & B\mathbb{Z}/2^{r-1} \end{array}$$

with

$$E_2^{*,*}(\pi) = \mathbb{Z}/2[\alpha, w_1, w_2, \ldots, w_k]$$

and

$$E_2^{*,*}(\pi_r) = \mathbb{Z}/2[\alpha, \beta, w_1, w_2, \ldots, w_k]/<\alpha^2 = 0>$$

if $r > 1$. The map Θ induces a monomorphism

$$\Theta^*: E_2^{even,*}(\pi_r) \longrightarrow E_2^{even,*}(\pi)$$

$$\mathbb{Z}/2[\beta, w_1, \ldots, w_k] \longrightarrow \mathbb{Z}/2[\alpha^2, w_1, \ldots, w_k]$$

with $\Theta^*(\beta) = \alpha^2$ and $\Theta^*(w_i) = w_i$. We therefore obtain the next result immediately from 3.3.7.

3.3.15 LEMMA. In $E_2^{*,*}(\pi_r)$ the following holds

$$d(\alpha) = d(\beta) = 0$$

$$d(w_i) = \begin{cases} 0 & \text{if } i+k \text{ is odd} \\ \beta w_{i-1} & \text{if } i+k \text{ is even and } i > 0. \end{cases}$$

A calculation similiar to that used to obtain Corollary 3.3.14 gives us the next result.

3.3.16 THEOREM. If $r > 1$ and k is even then $E_\infty^{*,*}(\pi_r)$ is the $\mathbb{Z}/2[\alpha, \beta, w_1, w_2^2, w_3, \ldots, w_k^2]/<\alpha^2 = 0>$ module on ρ_S, $S \in S_{2n}$ with relations $\beta w_{2i-1} = \beta \rho_S = 0$, $(S \neq \emptyset)$.

Theorem 3.3.5 then follows.

3.4 GENERATORS OF $N_*(B\Gamma)$

By using the results of the last three sections we can calculate the generators of $N_*(B\Gamma)$. For the case that Γ is an odd order finite abelian group the result is almost trivial.

3.4.1 LEMMA. *If Γ is a finite abelian group of odd order then $N_*(B\Gamma)$ is a free N_* module on one generator Γ.*

Thus there is an N_* module isomorphism $N_*(B\Gamma) \xrightarrow{\cong} N_*$.

For the case $\Gamma = BO_k$ it is more convenient to work with vector bundles having O_k as structure group rather than with singular manifolds in BO_k. If $J = (j(1), j(2), \ldots, j(k))$ is a k-tuple of non-negative integers with $j(1) \geq j(2) \geq \ldots \geq j(k)$ then let ξ_J be the bundle

$$\xi_{j(1)} \times \xi_{j(2)} \times \ldots \times \xi_{j(k)}$$

over $RP^J = RP^{j(1)} \times RP^{j(2)} \times \ldots \times RP^{j(k)}$ where $\xi_{j(i)}$ is the canonical line bundle over $RP^{j(i)}$. An alternative way of defining ξ_J is by

$$p_1^*(\xi_{j(1)}) \oplus p_2^*(\xi_{j(2)}) \oplus \ldots \oplus p_k^*(\xi_{j(k)})$$

where $p_i : RP^J \to RP^{j(i)}$ is projection on to the i-th factor.

3.4.2 LEMMA. *The bundles ξ_J, $J = (j(1), j(2), \ldots, j(k))$ with $j(1) \geq j(2) \geq \ldots \geq j(k) \geq 0$ are an N_* base for $N_*(BO_k)$.*

PROOF. Recall that the inclusion $(\mathbb{Z}/2)^k \subset O_k$ induces an inclusion

$$H^*(BO_k; \mathbb{Z}/2) = \mathbb{Z}/2[w_1, w_2, \ldots, w_k] \subset H^*(B(\mathbb{Z}/2)^k; \mathbb{Z}/2)$$
$$= \mathbb{Z}/2[t_1, t_2, \ldots, t_k]$$

with w_i the i-th elementary symmetric polynomial in

$t_1, t_2, \ldots, t_k$. If $J = (j(1), j(2), \ldots, j(k))$ with $j(1) \geq j(2) \geq \ldots \geq j(k) \geq 0$ then let S_J be the smallest symmetric (homogeneous) polynomial in $t_1, t_2, \ldots, t_k$ which contains the monomial $t_1^{j(1)} t_2^{j(2)} \ldots t_k^{j(k)}$. The set of such S_J form a base of $H^*(BO_k; \mathbb{Z}/2)$.

Let $f_J : RP^J \to BO_k$ be the classifying map for the bundle ξ_J and let σ_J be the fundamental $\mathbb{Z}/2$ homology class of RP^J then

$$\mu(RP^J, f_J) = (f_J)_* \sigma_J.$$

An easy calculation gives

$$\langle s_I, (f_J)_* \sigma_J \rangle = \langle f_J^* s_I, \sigma_J \rangle = \begin{cases} 1 & \text{if } I = J \\ 0 & \text{if } I \neq J \end{cases}$$

(see [Milnor and Stasheff] for some related, but different, calculations). Thus the set $\{(f_J)_* \sigma_J; J\}$ is the basis of $H_*(BO_k; \mathbb{Z}/2)$ dual to $\{s_J; J\}$ and the lemma is proved.

REMARK. $w_1(\xi_j) = y$ where $H^*(RP^j) = \mathbb{Z}/2[y]/\langle y^{j+1} = 0 \rangle$ and so $\langle (w_1)^i(\xi_j), \sigma_j \rangle = 1$ if and only if $i = j$.

The group $\mathbb{Z}/2$ acts on ξ_j by multiplication by -1 in each fibre thus we can form a line bundle

$$S^1 \times_{\mathbb{Z}/2} \xi_j$$

over $RP^1 \times RP^j$. The first Stiefel-Whitney class of this bundle is

$$w_1(S^1 \times_{\mathbb{Z}/2} \xi_j) = x+y$$

where $H^*(RP^1 \times RP^j) = \mathbb{Z}/2[x,y]/\langle x^2 = y^{j+1} = 0 \rangle$. (This can be easily seen by restricting the bundle to RP^1 and RP^j.) Thus if j is even then

$$\langle w_1^{j+1}(S^1 \times_{\mathbb{Z}/2} \xi_j), \sigma_1 \times \sigma_j \rangle = 1 \quad .$$

An alternative base for $N_*(BO_k)$ may therefore be obtained by replacing ξ_j, j odd, by $S^1 \times_{\mathbb{Z}/2} \xi_{j-1}$.

3.4.3 LEMMA. *The bundles* λ_J, $J = (j(1), j(2), \ldots, j(k))$ *with* $j(1) \geq j(2) \geq \ldots \geq j(k) \geq 0$ *are an* N_* *base for* $N_*(BO_k)$ *where* $\lambda_J = \lambda_{j(1)} \times \lambda_{j(2)} \times \ldots \times \lambda_{j(k)}$ *and*

$$\lambda_j = \begin{cases} \xi_j & \text{if } j \text{ is even} \\ S^1 \times_{\mathbb{Z}/2} \xi_{j-1} & \text{if } j \text{ is odd} \end{cases} \quad .$$

The result for $N_*(BU_\ell)$ is quite similiar. Let η_J be the bundle $\eta_{j(1)} \times \eta_{j(2)} \times \ldots \times \eta_{j(\ell)}$ over $CP^J = CP^{j(1)} \times CP^{j(2)} \times \ldots \times CP^{j(\ell)}$ where $\eta_{j(i)}$ is the canonical complex line bundle over $CP^{j(i)}$.

3.4.4 LEMMA. *The bundles* η_J, $J = (j(1), j(2), \ldots, j(\ell))$ *with* $j(1) \geq j(2) \geq \ldots \geq j(\ell) \geq 0$ *are an* N_* *base for* $N_*(BU_\ell)$.

The proof is similiar to that of Lemma 3.4.2, we use the inclusion $(S^1)^\ell \subseteq U_\ell$ which induces an inclusion

$$H^*(BU_\ell) = \mathbb{Z}/2[c_1, c_2, \ldots, c_\ell] \subset H^*(B(S^1)^\ell) = \mathbb{Z}/2[t_1, t_2, \ldots, t_\ell]$$

with $t_1, t_2, \ldots, t_\ell$ of degree 2.

Alternatively we could replace each of η_{2j+1} by the bundle

$$S^3 \times_{S^1} \eta_{2j}$$

over $CP^1 \times CP^{2j}$, where S^1 acts on S^3 by

$$t(z_1, z_2) = (tz_1, tz_2)$$

for $t \in S^1 \subset C$, $(z_1, z_2) \in S^3 \subset C^2$ and S^1 acts on η_{2j} by multiplication in each fibre ($=C$) of η_{2j}.

3.4.5 LEMMA. *The bundles ψ_J, $J = (j(1), j(2), \ldots, j(\ell))$ with $j(1) \geq j(2) \geq \ldots \geq j(\ell) \geq 0$ are an N_* base for $N_*(BU_\ell)$ where $\psi_J = \psi_{j(1)} \times \psi_{j(2)} \times \ldots \times \psi_{j(\ell)}$ and*

$$\psi_j = \begin{cases} \eta_j & \text{if } j \text{ is even} \\ S^3 \times_{S^1} \eta_{j-1} & \text{if } j \text{ is odd}. \end{cases}$$

Let us now consider the cyclic 2 groups $\mathbb{Z}/2^r$. If $r = 1$ then $H^*(B\mathbb{Z}/2) \cong \mathbb{Z}/2[\alpha]$. Thinking of elements of $N_*(B\mathbb{Z}/2)$ as manifolds with free involution we get the next result immediately from 3.4.2 and 3.4.3.

3.4.6 LEMMA. *(a) $\{S^n; n \geq 0$ and $\mathbb{Z}/2$ acting antipodally$\}$ is a free N_* base for $N_*(B\mathbb{Z}/2)$.*

(b) Alternatively the set
$$\{S^{2n}, S^1 \times_{\mathbb{Z}/2} S^{2n}; n \geq 0\}$$
is also a free N_* base for $N_*(B\mathbb{Z}/2)$.

For the case $r > 1$ we shall also write down two (different) bases for $N_*(B\mathbb{Z}/2^r)$.

3.4.7 LEMMA. *Each of the following*
  (a) $\{\mathbb{Z}/2^r \times_{\mathbb{Z}/2} S^{2n}, S^{2n+1}; n \geq 0\}$
  (b) $\{\mathbb{Z}/2^r \times_{\mathbb{Z}/2} S^{2n}, S^1 \times_{\mathbb{Z}/2} S^{2n}; n \geq 0\}$

is a free N_* base for $N_*(B\mathbb{Z}/2^r)$. The action of $\mathbb{Z}/2^r$ on $S^1 \subset C$ and on $S^{2n+1} \subset C^{2n}$ is induced by multiplication by $\exp(2\pi i/2^r)$.

PROOF. Consider the fibration
$$B\mathbb{Z}/2 \xrightarrow{i} B\mathbb{Z}/2^r \xrightarrow{\pi} B\mathbb{Z}/2^{r-1}.$$

Let $f: RP^{2n} \to B\mathbb{Z}/2$ classify the principle $\mathbb{Z}/2$ bundle S^{2n}. The map $if: RP^{2n} \to B\mathbb{Z}/2^r$ classifies the principal $\mathbb{Z}/2^r$ bundle $\mathbb{Z}/2^r \times_{\mathbb{Z}/2} S^{2n}$. This is not difficult to see - for example, let S^∞ denote the infinite sphere in C^∞. Since S^1 and all its subgroups act freely on S^∞ in the obvious way, we have $B\mathbb{Z}/2 = S^\infty/(\mathbb{Z}/2)$ and $B\mathbb{Z}/2^r = S^\infty/(\mathbb{Z}/2^r)$. It is then simple to verify that $i^*(E\mathbb{Z}/2^r) = \mathbb{Z}/2^r \times_{\mathbb{Z}/2} E\mathbb{Z}/2$ from which it follows that $(if)^*(E\mathbb{Z}/2^r) = \mathbb{Z}/2^r \times_{\mathbb{Z}/2} S^{2n}$.

We therefore have a commutative diagram of fibrations

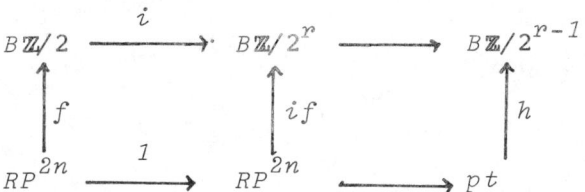

Recall that $H^*(B\mathbb{Z}/2^r) = \mathbb{Z}/2[\alpha,\beta]/\langle \alpha^2 = 0 \rangle$ with α coming from α in $H^*(B\mathbb{Z}/2^{r-1})$ and β restricting to α^2 in $H^*(B\mathbb{Z}/2)$. It is not difficult to verify that

$$\langle \alpha^j \beta^k, (if)_* \sigma \rangle = \begin{cases} 1 & \text{if } j = 0 \text{ and } k = 2n \\ 0 & \text{otherwise} \end{cases}$$

where σ is the fundamental homology class of RP^{2n}. In fact this situation is a simple case of a more general situation. Pull back the class $\alpha^j \beta^k$ to $H^*(RP^{2n})$ and call it $\bar{\alpha}\bar{\beta}$, i.e.,

$$\bar{\alpha} = (if)^*(\alpha^j), \quad \bar{\beta} = (if)^*(\beta^k)$$

then $\bar{\alpha}$ is a class coming from $h^*\alpha^j \in H^*(pt)$ and $\bar{\beta}$ restricts to the class $f^*\alpha^{2k} \in H^*(RP^{2n})$. Then we have

$$<\bar{\alpha}\bar{\beta}, \sigma> = <h^*\alpha^j, \sigma_{pt}><f^*\alpha^{2k}, \sigma>$$

which establishes the calculation above. More generally we have the following result.

3.4.8 LEMMA. *Suppose that*

$$Y \overset{i}{\hookrightarrow} X \times Y \overset{p}{\to} X$$

is a product fibration of manifolds and let $\bar{x} \in H^k(X \times Y)$, $\bar{y} \in H^\ell(X \times Y)$. *If* $\bar{x} = p^*x$ *for some* $\alpha \in H^k(X)$ *and if* $i^*\bar{y} = y \in H^\ell(Y)$ *then*

$$<\bar{x}\bar{y}, \sigma_{X \times Y}> = \begin{cases} <x, \sigma_X><y, \sigma_Y> & \text{if } k = \dim X \text{ and } \ell = \dim Y \\ 0 & \text{if } k+\ell \neq \dim(X \times Y) \text{ or} \\ & \text{if } k > \dim X \text{ (or } \ell < \dim Y). \end{cases}$$

PROOF. This is not difficult. We write elements of

$$H^*(X \times Y) \cong H^*(X) \otimes H^*(Y)$$

as pairs (x,y). Thus we have

$$\bar{x} = (x, 1)$$
$$\bar{y} = (1, y) + \sum_{i=1}^{\ell} (a_i, b_{\ell-i})$$

for some $a_i \in H^i(X)$, $b_{\ell-i} \in H^{\ell-i}(Y)$. So

$$<\bar{x}\bar{y}, \sigma_{X\times Y}> = <x,\sigma_X><y,\sigma_Y> + \sum_{i=1}^{\ell} <xa_i, \sigma_X><b_{\ell-i}, \sigma_Y>$$

and of course $xa_i \in H^{k+i}(X)$, $b_{\ell-i} \in H^{\ell-i}(Y)$, $i \geq 1$. The result follows immediately.

Returning to the proof of Lemma 3.4.7, since S^1 acts freely on S^∞ and $B\mathbb{Z}/2^r = S^\infty/(\mathbb{Z}/2^r)$ we see that $S^1/(\mathbb{Z}/2^r)$ ($\cong S^1$) also acts on $B\mathbb{Z}/2^r$. Define

$$g: S^1/(\mathbb{Z}/2^r) \times RP^{2n} \longrightarrow B\mathbb{Z}/2^r$$

$$(t,x) \longmapsto t(if(x))$$

where $if: RP^{2n} \to B\mathbb{Z}/2^r$ was given above. We have

$$g^*(E\mathbb{Z}/2^r) = S^1 \times_{\mathbb{Z}/2} S^{2n}$$

and a commutative diagram of fibrations

$$\begin{array}{ccccc} B\mathbb{Z}/2 & \longrightarrow & B\mathbb{Z}/2^r & \longrightarrow & B\mathbb{Z}/2^{r-1} \\ \uparrow f & & \uparrow g & & \uparrow \\ RP^{2n} & \longrightarrow & S^1/(\mathbb{Z}/2^r)\times RP^{2n} & \longrightarrow & S^1/(\mathbb{Z}/2^r) \end{array}$$

Using Lemma 3.4.7 we see that

$$<\alpha^j \beta^k, g_*\sigma> = \begin{cases} 1 & \text{if } j = 1 \text{ and } k = 2n \\ 0 & \text{otherwise} \end{cases}$$

(β is of degree 2 and α is of degree 1), where σ is the fundamental homology class of $S^1/(\mathbb{Z}/2^r) \times RP^{2n}$. Part (b) of the lemma follows immediately.

The fact that we may use S^{2n+1} in place of $S^1 \times_{\mathbb{Z}/2} S^{2n}$ in a base for $N_*(B\mathbb{Z}/2^r)$ follows from the fact that the principal bundle

$$S^{2n+1} \to S^{2n+1}/(\mathbb{Z}/2^r)$$

is $(2n+1)$ universal for principal $\mathbb{Z}/2^r$ bundles. This proves part (a) of Lemma 3.4.7.

The next groups that we have to consider are those of the form $(\mathbb{Z}/4 \times_{\mathbb{Z}/2} O_k)$. The result for k even is given first.

3.4.9 LEMMA. $N_*(B(\mathbb{Z}/4 \times_{\mathbb{Z}/2} O_{2n}))$ is generated as an N_* module by

$$\mathbb{Z}/4 \times_{\mathbb{Z}/2} \xi_J$$
$$S^1 \times_{\mathbb{Z}/2} \xi_J$$
$$S^n \times \eta_{2K}$$

where $J = (j_1, j_2, \ldots, j_{2n})$ with $j_1 \geq j_2 \geq \ldots \geq j_{2n} \geq 0$ and $2K = (2k_1, 2k_2, \ldots, 2k_n)$ with $k_1 \geq k_2 \geq \ldots \geq k_n \geq 0$. The group $\mathbb{Z}/4$ acts on S^n by multiplication by -1 and on each fibre of η_{2K} by multiplication by i.

Before proving the result we state another version of it.

3.4.10 LEMMA. $N_*(B(\mathbb{Z}/4 \times_{\mathbb{Z}/2} O_{2n}))$ is generated as an N_* module by

$$\mathbb{Z}/4 \times_{\mathbb{Z}/2} \xi_J$$
$$S^1 \times_{\mathbb{Z}/2} \xi_J$$
$$S^{2i} \times \eta_{2K}$$
$$S^1 \times_{\mathbb{Z}/4} S^{2i} \times \eta_{2K} .$$

PROOF. The idea is to use the fibration

$$BO_{2n} \to B\Gamma \to B\mathbb{Z}/2$$

where $\Gamma = \mathbb{Z}/4 \times_{\mathbb{Z}/2} O_{2n}$. We also have the fibrations

(i) $\xi_J \to \mathbb{Z}/4 \times_{\mathbb{Z}/2} \xi_J \to \mathbb{Z}/2$

(ii) $\xi_J \to S^1 \times_{\mathbb{Z}/2} \xi_J \to S^1/(\mathbb{Z}/2)$

(iii) $\eta_{2K} \to S^n \times \eta_{2K} \to S^n$

(iv) $\eta_{2K} \to S^{2i} \times \eta_{2K} \to S^{2i}$

(v) $\eta_{2K} \to S^1 \times_{\mathbb{Z}/4} S^{2i} \times \eta_{2K} \to S^1/(\mathbb{Z}/2) \times_{\mathbb{Z}/2} S^{2i}$.

Denote any of these by

$$F' \to E' \to B'$$

where F', E', B' are respectively, principal $O_{2n}, \Gamma, \mathbb{Z}/2$ bundles. Denote by

$$F \to E \to B$$

the fibration of the appropriate bases spaces, i.e., F is the base space of the principal O_{2n} bundle F' etc. In each case there is a commutative diagram of fibrations

$$\begin{array}{ccc} BO_{2n} & \xrightarrow{i} B\Gamma \longrightarrow & B\mathbb{Z}/2 \\ \uparrow f & \uparrow g & \uparrow h \\ F & \longrightarrow E \longrightarrow & B \end{array}$$

where f, g, h classify, respectively, the principal bundles F', E', B'. For case (i) this is trivial. For case (ii) we need an S^1 action on $B\Gamma$. Since $\Gamma = \mathbb{Z}/4 \times_{\mathbb{Z}/2} O_{2n}$ is a subgroup of

$S^1 \times_{\mathbb{Z}/2} O_{2n}$ we may take $E\Gamma$ as $E(S^1 \times_{\mathbb{Z}/2} O_{2n})$ which has a free $S^1/(\mathbb{Z}/2)$ action and so $B\Gamma$ has a free $S^1/(\mathbb{Z}/4) \cong S^1$ action. Defining g as follows

$$g : S^1/(\mathbb{Z}/4) \times RP^J \to B\Gamma$$

$$(t,x) \to t(if(x))$$

(see also the proof of 3.4.7) we then see that

$$g^*(E\Gamma) = S^1 \times_{\mathbb{Z}/2} \xi_J \;.$$

For cases (iii), (iv) and (v) we first establish a commutative diagram of fibrations

$$\begin{array}{ccccc} BU_n & \longrightarrow & B(\mathbb{Z}/4 \times_{\mathbb{Z}/2} U_n) & \longrightarrow & B\mathbb{Z}/2 \\ \uparrow & & \uparrow & & \uparrow \\ F & \longrightarrow & E & \longrightarrow & B \end{array}$$

which is easy because $\mathbb{Z}/4 \times_{\mathbb{Z}/2} U_n \cong \mathbb{Z}/2 \times U_n$. The natural inclusions $U_n \hookrightarrow O_{2n}$, $\mathbb{Z}/4 \times_{\mathbb{Z}/2} U_n \hookrightarrow \mathbb{Z}/4 \times_{\mathbb{Z}/2} O_{2n}$ lead then to the required commutative diagram. The proof of Lemma 3.4.9 and 3.4.10 is then completed by using Lemma 3.4.8 and a simple calculation.

3.4.11 EXERCISE. Write down an N_* basis for $N_*(B(\mathbb{Z}/4 \times_{\mathbb{Z}/2} O_k))$, where $k = 2n$.

The result for k odd is given next.

3.4.12 LEMMA. (i) $N_*(B(\mathbb{Z}/4 \times_{\mathbb{Z}/2} O_{2n+1}))$ *is generated as an* N_* *module by*

$$\mathbb{Z}/4 \times_{\mathbb{Z}/2} \xi_J$$

$$S^1 \times_{\mathbb{Z}/2} \xi_J$$

where $J = (j_1, j_2, \ldots, j_{2n+1})$ and $j_1 \geq j_2 \geq \ldots \geq j_{2n+1} \geq 0$.

(ii) An N_* basis for $N_*(B(\mathbb{Z}/4 \times_{\mathbb{Z}/2} O_1))$ is given by

$$\mathbb{Z}/4 \times_{\mathbb{Z}/2} \xi_{2j}$$
$$S^1 \times_{\mathbb{Z}/2} \xi_{2j}$$

where $j \geq 0$.

PROOF. We know that the cohomology of $B(\mathbb{Z}/4 \times_{\mathbb{Z}/2} O_{2n+1})$ = cohomology of $B\mathbb{Z}/4 \times BSO_{2n+1}$ has a basis

$$\alpha^i \beta^j w_2^{k(2)} \ldots w_{2n+1}^{k(2n+1)} \quad i = 0, 1, \, j \geq 0, \, k_2 \geq k_3 \geq \ldots \geq k_{2n+1}$$

where α, β come from $H^*(B\mathbb{Z}/4)$ and $w_i \in H^i(BSO_{2n+1})$. Consider the fibration

$$BO_{2n+1} \to B(\mathbb{Z}/4 \times_{\mathbb{Z}/2} O_{2n+1}) \to B\mathbb{Z}/2 .$$

Then we may consider α coming from $H^*(B\mathbb{Z}/2)$ and $\beta, w_2, \ldots, w_{2n+1}$ restricting respectively to $w_1^2, w_2, \ldots, w_{2n+1}$ in $H^*(BO_{2n+1})$ [ignoring decomposables]. It is not difficult to show that there are commutative diagrams of fibrations

$$\begin{array}{ccccc}
BO_{2n+1} & \longrightarrow & B(\mathbb{Z}/4 \times_{\mathbb{Z}/2} O_{2n+1}) & \longrightarrow & B\mathbb{Z}/2 \\
\uparrow f & & \uparrow g & & \uparrow h \\
RP^J & \longrightarrow & X/(\mathbb{Z}/4) \times RP^J & \longrightarrow & X/(\mathbb{Z}/4)
\end{array}$$

where X is either $\mathbb{Z}/4$ or S^1 and f classifies ξ_J, g classifies

$X \times_{\mathbb{Z}/2} \xi_J$ and h classifies $X/(\mathbb{Z}/2)$. The result follows after a simple calculation using Lemma 3.4.8.

3.4.13 EXERCISE. The homomorphism

$$O_{2n+1} \to SO_{2n+1}$$
$$\alpha \to \alpha \det \alpha$$

induces a homomorphism

$$H^*(BSO_{2n+1}; \mathbb{Z}/2) \to H^*(BO_{2n+1}; \mathbb{Z}/2) .$$

Show that under this homomorphism

$$w_i \to \sum_{j=0}^{i} \binom{n-i+j}{j} w_1^j w_{i-j} .$$

In the same way as the above we can now prove the more general result.

3.4.14 LEMMA. *If* $r > 1$ *then* $N_*(B(\mathbb{Z}/2^r \times_{\mathbb{Z}/2} O_k)$ *is generated as an* N_* *module by*

$$\mathbb{Z}/2^r \times_{\mathbb{Z}/2} \xi_J$$
$$S^1 \times_{\mathbb{Z}/2} \xi_J$$
$$\left. \begin{array}{l} (\mathbb{Z}/2^r \times_{\mathbb{Z}/4} S^{2i}) \times \eta_{2K} \\ (S^1 \times_{\mathbb{Z}/4} S^{2i}) \times \eta_{2K} \end{array} \right\} \text{ if } k \text{ is even}$$

In the case $k = 1$

$$\mathbb{Z}/2^r \times_{\mathbb{Z}/2} \xi_{2j}$$
$$S^1 \times_{\mathbb{Z}/2} \xi_{2j}$$

form an N_* *basis.*

3.4.15 COROLLARY. $N_*(B(\mathbb{Z}/2^r \times O_k)/(\mathbb{Z}/2^s))$ is generated as an N_* module by

$$\mathbb{Z}/2^r \times_{\mathbb{Z}/2^s} \xi_J$$

$$S^1 \times_{\mathbb{Z}/2^s} \xi_J$$

$$\left.\begin{array}{l} (\mathbb{Z}/2^r \times_{\mathbb{Z}/2^{s+1}} S^{2i}) \times \eta_{2K} \\ (S^1 \times_{\mathbb{Z}/2^{s+1}} S^{2i}) \times \eta_{2K} \end{array}\right\} \quad \text{if } r > s \text{ and } k \text{ is even.}$$

In the case $k = 1$

$$\mathbb{Z}/2^r \times_{\mathbb{Z}/2^s} \xi_{2j} \; ,$$

$$S^1 \times_{\mathbb{Z}/2^s} \xi_{2j}$$

form an N_* basis.

The subgroups $\mathbb{Z}/2^s$, $\mathbb{Z}/2^{s+1}$ act via their natural inclusions on $\mathbb{Z}/2^r$ and S^1. These subgroups act on ξ_J and S^{2i} via their projection onto $\mathbb{Z}/2$.

The proof of the next result is obvious and will not be given.

3.4.16 LEMMA. If $\{A_i; i \in I\}$, $\{B_j; j \in J\}$ are respectively an N_* base (or generating set) of $N_*(BG_1)$, $N_*(BG_2)$ then $\{A_i B_j; (i,j) \in I \times J\}$ is an N_* base (or generating set) of $N_*(B(G_1 \times G_2))$.

For example, an N_* base for $N_*(BO_1 \times BU_1)$ is given by

$$\xi_i \times \eta_j \qquad i \geq 0, \; j \geq 0 \; .$$

In chapter 4 we shall have need of a different basis. This is given in the next lemma.

3.4.17 LEMMA. *An N_* base for $N_*(BO_1 \times BU_1)$ is given by*

$$\xi_a \times \eta_{2b}$$

$$(\xi_{a+2} \times \eta_{2b}) + (\xi_{a+2} \times (\eta_{2b} \otimes \xi_{a+2}))$$

for $a, b \geq 0$.

Recall that $\xi_{a+2} \times \eta_{2b}$ is the same as

$$p_1^*(\xi_{a+2}) \oplus p_2^*(\eta_{2b})$$

where $p_1: RP^{a+2} \times CP^{2b} \to RP^{a+2}$, $p_2: RP^{a+2} \times CP^{2b} \to CP^{2b}$ are the canonical projections. By $\xi_{a+2} \times (\eta_{2b} \otimes \xi_{a+2})$ we mean

$$p_1^*(\xi_{a+2}) \oplus (p_2^*(\eta_{2b}) \otimes p_1^*(\xi_{a+2})) .$$

PROOF OF LEMMA. The Stiefel-Whitney classes of the bundles in question are given by

$$w(\xi_a \times \eta_{2b}) = (1+x)(1+z)$$
$$w(\xi_a \times (\eta_{2b} \otimes \xi_a)) = (1+x)(1+x^2+z)$$

where $H^*(RP^a) = (\mathbb{Z}/2)[x]/x^{a+1}$ and $H^*(CP^{2b}) = (\mathbb{Z}/2)[z]/z^{2b+1}$. The cohomology of $BO_1 \times BU_1$ is $(\mathbb{Z}/2)[s,t]$ where s is one dimensional and is the universal first Stiefel-Whitney class while t is two dimensional and is the universal first Chern class reduced mod 2. A simple calculation shows that

$$s^k t^\ell (\xi_a \times \eta_{2b}) = \begin{cases} 1 & \text{if } k = a \text{ and } \ell = 2b \\ 0 & \text{otherwise} \end{cases}$$

and

$$s^k t^\ell ((\xi_{a+2} \times \eta_{2b}) + (\xi_{a+2} \times (\eta_{2b} \otimes \xi_{a+2}))) = \begin{cases} 1 & \text{if } k = a \text{ and } \ell = 2b+1 \\ 0 & \text{if } k > a \text{ or if } \ell < 2b \end{cases}$$

where by $s^k t^\ell(E)$ we mean $<f^*(s^k t^\ell), \sigma_X>$, f being the classifying map $f: X \to BO_1 \times BU_1$ of the bundle E over X. This completes the proof of the lemma.

3.4.18 EXERCISE. Prove that in $N_*(BO_1 \times BU_1)$ we have the following equality.

$$(\xi_{a+2} \times \eta_{2b}) + (\xi_{a+2} \times (\eta_{2b} \otimes \xi_{a+2})) =$$
$$(\xi_a \times \eta_{2b+1}) + \sum_{m=2}^{(a+2)/2} \binom{2b+m}{m} (\xi_{a+2-2m} \times \eta_{2b+m}) + decomposables$$

where by $decomposables$ we mean terms of the form $M \times \xi_i \times \eta_j$ with $\dim M + i + 2j = a+2+4b$ and $\dim M > 0$.

Lemma 3.4.17 may be generalised.

3.4.19 LEMMA. *An N_* base for $N_*(BO_1 \times BU_\ell)$ is given by*

$$\xi_a \times \eta_{2J}$$
$$(\xi_{a+2} \times \eta_{2b} \times \eta_I) + (\xi_{a+2} \times (\eta_{2b} \otimes \xi_{a+2}) \times \eta_I)$$

where $a, b \geq 0$, $2J = (2j_1, 2j_2, \ldots, 2j_\ell)$ *with* $j_1 \geq j_2 \geq \ldots \geq j_\ell \geq 0$, $I = (i_2, i_3, \ldots, i_\ell)$ *with* $i_2 \geq i_3 \geq \ldots \geq i_\ell \geq 0$ *and* $2b+1 \geq \max\{i_j; i_j \text{ odd}\}$.

The result is proved either by using Exercise 3.4.18 or generalising the proof of Lemma 3.4.17. The details are left for the reader.

3.5. CALCULATION OF $SK_*(B\Gamma)$

We shall use results from the last section together with those from chapter 2 to obtain the next result.

3.5.1 THEOREM. *If Γ is one of the following groups:*

a finite abelian group, O_k, U_ℓ, $\mathbb{Z}/2^r \times_{\mathbb{Z}/2^s} O_k$ $(r > s)$,

or products of these groups then the natural homomorphism

$$SK_*(B\Gamma) \to SK_*$$

is an isomorphism.

PROOF. For each of the groups Γ mentioned in the theorem, we see from the last section that we can find generators of $N_*(B\Gamma)$ that are fibre bundles over products of real or complex projective spaces. The result follows from Lemma 2.3.3 and Theorem 2.4.1.

3.5.2 COROLLARY. *The SK_* homomorphism*

$$SK_* \to SK_*^G[H;U]$$

given by

$$M \to M \times (G \times_H U)$$

is an SK_* isomorphism.

3.6 HISTORICAL NOTE

The calculation of $H^*(B(\mathbb{Z}/2^r \times_{\mathbb{Z}/2} O_k))$ was first obtained in [Beem,1]. The method used in [Beem, 1] is also the Serre Spectral Sequence although the details are different. The proof presented here is perhaps more pedestrian.

The generators of $N_*(B\Gamma)$ for certain groups $(\mathbb{Z}/2^r, O_k, U_\ell)$ is classical. The result concerning $SK_*(B\Gamma)$ for *some* of the groups listed in 3.5.1 appears in [Karras, Kreck, Neumann and Ossa].

4 Generators of the equivariant bordism groups

4.1 GENERATORS FOR G BORDISM, G ODD ORDER

For the case of an odd order group G it is easier to write down *multiplicative* generators of N_*^G. Let η_j denote the canonical complex line bundle over CP^j. If V is an irreducible G module then by $\eta_j \otimes V$ we mean the G vector bundle over CP^j given by the tensor product of η_j (with trivial G action) and the trivial vector bundle $V \times CP^j$ with G action induced by the action of G on V.

4.1.1 THEOREM. *If G is a finite abelian group of odd order then N_*^G is multiplicatively generated over N_* by the following G manifolds:*

$$\{G/H; H \subseteq G\} \cup$$
$$\{RP(R \times (\eta_j \otimes V)); j \geq 0, V \text{ irreducible } G \text{ module}\}.$$

The proof is not difficult. By Corollaries 1.7.4 and 1.7.5 we have

$$N_*^G \cong \bigoplus_{[H;U] \in St(G)} N_*^G[H;U]$$

$$\cong \bigoplus_{[H;U] \in St(G)} N_{*-|U|}(B\Gamma(H;U))$$

where $|U| = \dim U$. We know that $\Gamma(H;U)$ takes the form

$$G/H \times U_{j(1)} \times U_{j(2)} \times \ldots \times U_{j(k)}$$

for some k-tuple $(j_1, j_2, \ldots, j_k)$. The generators of $N_*(B(G/H))$ and $N_*(BU_{j(i)})$ were calculated in the last chapter. We see that the generators of $N_*^G[H;U]$ are products of G/H with products of terms of the form

$$\eta_j \otimes V \quad j \geq 0, \; V \text{ irreducible } G \text{ module}, \; V|H \subseteq U.$$

Note that if V is an H module then it is also a G module by 3.2.1 and so $G \times_H (\eta_j \otimes V)$ is isomorphic to $(G/H) \times (\eta_j \otimes V)$.

Instead of using the splittings

$$N_*^G[H;U] \longrightarrow N_*^G$$

$$E \longmapsto RP(R \times E)$$

we use the splittings which on the generators of $N_*^G[H;U]$ are given by

$$G/H \to G/H$$

$$\eta_j \otimes V \to RP(R \times (\eta_j \otimes V)).$$

The details of the fact that this defines an N_* homomorphism and a splitting is left for the reader. The theorem follows immediately.

4.1.3 EXERCISE. Write down an N_* *basis* for N_*^G when G is a finite abelian group of odd order.

4.2 GENERATORS FOR $\mathbb{Z}/2$ BORDISM

Let ξ_m denote the canonical line bundle over RP^m. Throughout this section it will be assumed that the group $\mathbb{Z}/2$ acts on ξ_m via multiplication by -1 in each fibre of ξ_m. We define, as before, ξ_J to be the bundle $\xi_{j(1)} \times \xi_{j(2)} \times \ldots \times \xi_{j(k)}$ over $RP^J = RP^{j(1)} \times RP^{j(2)} \times \ldots \times RP^{j(k)}$ where $J = (j(1), j(2), \ldots, j(k))$ is a k-tuple of non-negative integers. If $J = \emptyset$ then $\xi_\emptyset$ = point.

4.2.1 THEOREM. $N_*^{\mathbb{Z}/2}$ *is freely generated as an* N_* *module by the following* $\mathbb{Z}/2$ *manifolds:*

$$\{RP(R \times \xi_J); \ J = (j_1, j_2, \ldots, j_k), \ j_1 \geq j_2 \geq \ldots \geq j_k \geq 0, \ k \geq 0, \ k \neq 1\}$$

The rest of this section will be devoted to proving this and related results. The $\mathbb{Z}/2$-slice types are of the form $[\mathbb{Z}/2; \tilde{R}^j]$, $j \geq 0$ and $[1; 0]$ where $\tilde{R}$ denotes the real numbers with $\mathbb{Z}/2$ acting via multiplication by -1. Let σ_j denote the slice type $[\mathbb{Z}/2; \tilde{R}^j]$ if $j \geq 0$ and let σ_{-1} denote $[1; 0]$. Define families F_j, $j \geq -1$, of $\mathbb{Z}/2$-slice types by

$$F_j = \{\sigma_{-1}, \sigma_0, \ldots, \sigma_j\} \ .$$

We therefore have

$$N_n^{\mathbb{Z}/2}[F_N] = N_n^{\mathbb{Z}/2} \quad \text{if } N \geq n \ .$$

From Chapter 1 we have the following exact sequence

$$\ldots \to N_n^G[F_{j-1}] \to N_n^G[F_j] \xrightarrow{\nu_j} N_n^G[\sigma_j] \xrightarrow{\partial_j} N_{n-1}^G[F_{j-1}] \to \ldots$$

where $G = \mathbb{Z}/2$. If $E \in N_n^G[\sigma_j]$ then it is easy to check that $RP(R \times E)$ is a $\mathbb{Z}/2$ manifold of type F_j. Indeed the slice types of $RP(R \times E)$ are precisely σ_{-1}, σ_j and σ_1 if $j \geq 1$ and σ_j otherwise. We can either use Corollary 1.7.3 or a direct argument to show that the N_* homomorphism

$$q_j : N_n^G [\sigma_j] \to N_n^G [F_j]$$

$$E \mapsto RP(R \times E)$$

splits the above exact sequence if $j \neq 1$.

4.2.2 COROLLARY.

$$N_n^{\mathbb{Z}/2} \cong \bigoplus_{j=2}^{n} N_n^{\mathbb{Z}/2} [\sigma_j] \oplus N_n^{\mathbb{Z}/2} [F_1] .$$

We therefore need to calculate $N_n^{\mathbb{Z}/2} [F_1]$. Consider the homomorphisms

$$N_n^G [\sigma_1] \xrightarrow{\partial_1} N_{n-1}^G [F_0]$$

with q_{-1} upward and ν_{-1} downward to

$$N_{n-1}^G [F_{-1}]$$

4.2.3 LEMMA. $\nu_{-1} \partial_1$ *is an isomorphism.*

PROOF. $N_{n-1}^G [F_{-1}]$ consists of free $(n-1)$ dimensional $\mathbb{Z}/2$ manifolds while $N_n^G [\sigma_1]$ consists of line bundles over $(n-1)$ dimensional manifolds with $\mathbb{Z}/2$ acting in the fibre only. An inverse to $\nu_{-1} \partial_1$ is given by associating to each free $\mathbb{Z}/2$ manifold M the line bundle associated to the double covering $M \to M/(\mathbb{Z}/2)$.

It therefore follows that ∂_1 is injective and hence the next lemma follows immediately.

4.2.4 LEMMA.

$$N_n^{\mathbb{Z}/2} \cong \bigoplus_{\substack{j=0 \\ j \neq 1}}^{n} N_n^{\mathbb{Z}/2}[\sigma_j]$$

$$\cong \bigoplus_{\substack{j=0 \\ j \neq 1}}^{n} N_{n-j}(BO_j) .$$

Furthermore, the isomorphism $q: \oplus N_n^{\mathbb{Z}/2}[\sigma_j] \to N_n^{\mathbb{Z}/2}$ is given by $q(E_0, E_2, E_3, \ldots, E_n) = \sum q_j(E_j)$ where $q_j(E_j) = RP(R \times E_j)$ for $E_j \in N_n^{\mathbb{Z}/2}[\sigma_j]$.

The main result now follows from knowledge of the generators of $N_{n-j}(BO_j)$ and the explicit isomorphism in the above lemma.

REMARK. $N_n^{\mathbb{Z}/2}[\sigma_{-1}]$ and $N_n^{\mathbb{Z}/2}[\sigma_1]$ do not appear in Lemma 4.2.4. In fact if M is a $\mathbb{Z}/2$ manifold which is either free or has a codimension one only fixed point set then M is a $\mathbb{Z}/2$ boundary. In each case the quotient $M/(\mathbb{Z}/2)$ is a manifold and also the mapping cylinder of $M \to M/(\mathbb{Z}/2)$ is a manifold whose boundary is precisely M.

4.3 GENERATORS OF $\mathbb{Z}/4$ BORDISM

The main result of this section gives the generators of $\mathbb{Z}/4$ bordism. We leave the general case of $\mathbb{Z}/2^r$ bordism to the next section. If I is a sequence of non-negative integers then we denote the length of I by $l(I)$.

4.3.1 THEOREM. $N_*^{\mathbb{Z}/4}$ is generated as an N_* module by the *following* $\mathbb{Z}/4$ *manifolds:*

$\mathbb{Z}/4 \times_{\mathbb{Z}/2} RP(R \times \xi_I)$ $\qquad$ ($l(I) \geq 2$)

$S^1 \times_{\mathbb{Z}/2} RP(R \times \xi_I)$ $\qquad$ ($l(I) \geq 2$)

$RP(R \times \xi_I \times \eta_J)$ $\qquad$ ($l(I) \geq 0$, $l(I) \neq 1$, $l(J) \geq 0$)

$RP(R \times \xi_{a+2} \times \eta_{2b}) \times RP(R \times \eta_J)$ $\qquad$ ($l(J) \geq 0$)

where I and J are non-increasing sequences of non-negative integers while a and b are non-negative integers.

The action of the generator of $\mathbb{Z}/2$ on ξ_I is given by multiplication by -1 in each fibre of ξ_I and on $\mathbb{Z}/4$, S^1 the action is via the inclusion. The quotients

$\mathbb{Z}/4 \times_{\mathbb{Z}/2} RP(R \times \xi_I)$, $S^1 \times_{\mathbb{Z}/2} RP(R \times \xi_I)$

are therefore defined. We let $\mathbb{Z}/4$ act on these manifolds by the action of $\mathbb{Z}/4$ on the first factor, i.e., on $\mathbb{Z}/4$, S^1 respectively. The generator of $\mathbb{Z}/4$ acts on the manifolds

$RP(R \times \xi_I \times \eta_J)$

via the action induced by multiplication by -1 in each fibre of ξ_I and multiplication by $i = \sqrt{-1}$ in each fibre of η_J.

To prove the theorem we need to know the $\mathbb{Z}/4$-slice types. The non-trivial irreducible $\mathbb{Z}/4$ modules are

$\tilde{R}, \tilde{C}$

where $\mathbb{Z}/4$ (and $\mathbb{Z}/2$) acts on $\tilde{R}$ as ± 1 while $\mathbb{Z}/4$ acts on $\tilde{C}$ by multiplication by $i = \sqrt{-1}$. The $\mathbb{Z}/4$-slice types are therefore

$$\sigma_{-1} = [1;0]$$
$$\sigma_j = [\mathbb{Z}/2; \tilde{R}^j] \quad j \geq 0$$
$$\sigma_{k,l} = [\mathbb{Z}/4; \tilde{R}^k \times \tilde{C}^l] \quad k,l \geq 0.$$

We order the $\mathbb{Z}/4$ slice types by the following rules:

(a) If $k+2l > m+2n$ then $\sigma_{k,l} > \sigma_{m,n}$.
(b) If $k+2l = m+2n$ and $k > m$ then $\sigma_{k,l} > \sigma_{m,n}$.
(c) If $k+2l > m$ then $\sigma_{k,l} > \sigma_m$.
(d) If $k+2l = m$ then $\sigma_m > \sigma_{k,l}$.
(e) $\sigma_{k,l}, \sigma_k > \sigma_{-1}$ for all $k, l \geq 0$.

In fact this ordering of the $\mathbb{Z}/4$-slice types is similiar to that described in section 1.7 for the case of an odd order group. In other words we order first by the *dimension* of the $\mathbb{Z}/4$-slice types (dim $\sigma_{k,l} = k+2l$, dim $\sigma_k = k$). If the dimensions agree then we look at the isotropy subgroup of the $\mathbb{Z}/4$-slice type ($\mathbb{Z}/4 < \mathbb{Z}/2 < 1$). Finally, if the dimensions and isotropy subgroups are the same then this occurs non-trivially only with $\mathbb{Z}/4$ as isotropy subgroup. In this case we use the lexicographical ordering on the $\mathbb{Z}/4$ modules induced by saying that $\tilde{R} < \tilde{C}$. (Note that in the notation of section 1.6: $V_{\mathbb{Z}/4} = R$, $V_{\mathbb{Z}/2} = \tilde{R}$, $V_1 = \tilde{C}$.)

Observe that

(i) $\sigma_{k,l} > \sigma_{1,l}$ if $k > 1$, and
(ii) $\sigma_k > \sigma_1$ if $k > 1$.

Rename the $\mathbb{Z}/4$-slice types as

$\rho_0, \rho_1, \rho_2, \ldots$

with the condition that if $\rho_i < \rho_j$ (with respect to the given ordering) then $i < j$. Let F_j be the collection of $\mathbb{Z}/4$-slice types given by

$$F_j = \{\rho_0, \rho_1, \rho_2, \ldots, \rho_j\}.$$

The ordering on the $\mathbb{Z}/4$-slice types ensures that F_j is a family of $\mathbb{Z}/4$-slice types. In fact the relevant conditions for a family are

$$\sigma_{k,l} > \sigma_{2l}; \quad \sigma_{k,l}, \sigma_k > \sigma_{-1}.$$

If E is a G vector bundle of type $\sigma_{k,l}$ then $q(E) = RP(R \times E)$ has slice types

(i) $\sigma_{k,l}, \sigma_{2l}, \sigma_{1,l}, \sigma_1, \sigma_{-1}$ if $k, l \geq 1$
(ii) $\sigma_{0,l}, \sigma_{2l}, \sigma_1, \sigma_{-1}$ if $k = 0$, $l \geq 1$, and
(iii) $\sigma_{0,0}, \sigma_0$ if $k = l = 0$

(see section 1.7). If E is a G vector bundle of type σ_k then $q(E)$ has slice types

(i) $\sigma_k, \sigma_1, \sigma_{-1}$ if $k \geq 1$
(ii) σ_k otherwise.

Thus the observations $\sigma_{k,l} > \sigma_{1,l}, \sigma_k > \sigma_1$, if $k > 1$ and $\sigma_{k,l}, \sigma_k > \sigma_{-1}$ show that if E is a vector bundle of type ρ_j then $q(E)$ is a $\mathbb{Z}/4$ manifold of type F_j. We therefore deduce the next result from Corollary 1.7.3.

4.3.2 LEMMA. *The following sequence, for* $G = \mathbb{Z}/4$,

$$0 \to N_n^G[F_{j-1}] \xrightarrow{i} N_n^G[F_j] \xrightarrow{\nu} N_n^G[\rho_j] \to 0$$

is a split short exact sequence if $\rho_j \neq \sigma_{1,l}$ *or* σ_1. *A split-*

ting homomorphism q: $N_n^G[\rho_j] \to N_n^G[F_j]$ is given by $q(E) = RP(R \times E)$ for $E \in N_n^G[\rho_j]$.

For the case $\rho_j = \sigma_{2l}$ an alternative splitting homomorphism is given in Remark 4.3.11.

4.3.3 LEMMA. If $G = \mathbb{Z}/4$ then N_n^G is isomorphic to the direct sum of the following:

$$\bigoplus_{\substack{k \text{ odd} \\ k \neq 1}} N_n^G[\sigma_k]$$

$$\bigoplus_{\substack{k, l \\ k \neq 1}} N_n^G[\sigma_{k,l}]$$

$$\bigoplus_{l} (\ker(\nu\partial: N_n^G[\sigma_{1,l}] \to N_{n-1}^G[\sigma_{2l}]))$$

$$\bigoplus_{l} N_n^G[\sigma_{2l}]/(\operatorname{im}(\nu\partial: N_{n+1}^G[\sigma_{1,l}] \to N_n^G[\sigma_{2l}]))$$

where if $E \in N_n^G[\sigma_{1,l}]$ then $\nu\partial(E) = \nu_{2l} S(E)$. Furthermore, an isomorphism from the direct sum to N_n^G is given by

$$(\bigoplus_{\substack{k \text{ odd} \\ k \neq 1}} E_k) \oplus (\bigoplus_{\substack{k,l \\ k \neq 1}} E_{k,l}) \oplus (\bigoplus_{l} E_{1,l}) \oplus (\bigoplus_{l} E_{2l}) \longmapsto$$

$$\sum_{\substack{k \\ k \neq 1}} \phi_k(E_k) + \sum_{k,l} \phi_{k,l}(E_{k,l})$$

with $\phi_k(E_k) = RP(R \times E_k)$ and $\phi_{k,l}(E_{k,l}) = RP(R \times E_{k,l})$ if $k \neq 1$.

The proof of this theorem is by induction on the families $F_0, F_1, F_2, \ldots$ of G slice types. We shall state the result in a slightly different way, but first define subsets A_j, B_j of F_j inductively as follows:

$$A_0 = \{\rho_0\} \ (= \sigma_{-1}), \quad B_0 = \emptyset.$$

Suppose that A_{j-1} and B_{j-1} are defined. There are three possibilities for ρ_j:

(i) $\rho_j \neq \sigma_1, \sigma_{1,l}$

(ii) $\rho_j = \sigma_1$

(iii) $\rho_j = \sigma_{1,l}$.

(i) If $\rho_j \neq \sigma_1, \sigma_{1,l}$ then define

$$A_j = A_{j-1} \cup \{\rho_j\}$$
$$B_j = B_{j-1} .$$

(ii) If $\rho_j = \sigma_1$ then define

$$A_j = A_{j-1} - \{\sigma_{-1}\}$$
$$B_j = B_{j-1} \cup \{\sigma_1\}$$

and set $r(\sigma_1) = \sigma_{-1}$.

(iii) If $\rho_j = \sigma_{1,l}$ then define

$$A_j = A_{j-1} - \{\sigma_{2l}\}$$
$$B_j = B_{j-1} \cup \{\sigma_{1,l}\}$$

and set $r(\sigma_{1,l}) = \sigma_{2l}$.

We leave it for the reader to check that the definitions make sense and also that

$$A_j \cup B_j \cup \{r(\rho); \rho \in B_j\} = F_j .$$

4.3.4 LEMMA. *If $G = \mathbb{Z}/4$ then $N_n^G[F_j]$ is isomorphic to the direct sum of the following:*

$$\bigoplus_{\rho \in A_j} N_n^G[\rho]$$

$$\bigoplus_{\rho \in B_j} (\ker(\nu\partial : N_n^G[\rho] \to N_{n-1}^G[r(\rho)]))$$

$$\bigoplus_{\rho \in B_j} N_n^G[r(\rho)]/(\operatorname{im}(\nu\partial : N_{n+1}^G[\rho] \to N_n^G[r(\rho)])).$$

Furthermore, the isomorphism from the direct sum to $N_n^G[F_j]$, when restricted to the first and last terms is given by $E \mapsto RP(R \times E)$.

The reader should have no problems in obtaining Lemma 4.3.3 from 4.3.4 after noting that

$$\nu\partial : N_n^G[\sigma_1] \to N_{n-1}^G[\sigma_{-1}]$$

is an isomorphism.

We shall now prove Lemma 4.3.4 by induction on j. If $j = 0$ then

$$N_n^G[F_0] \cong N_n^G[\rho_0]$$

which agrees with the statement in Lemma 4.3.4 since $A_0 = \{\rho_0\}$ and $B_0 = \emptyset$. Suppose that the result in question holds for $j-1$ ($j \geq 1$), we have the following exact sequence

$$\ldots \to N_n^G[F_{j-1}] \to N_n^G[F_j] \xrightarrow{\nu_j} N_n^G[\rho_j] \xrightarrow{\partial_j} N_{n-1}^G[F_{j-1}] \to \ldots \;.$$

If $\rho_j \neq \sigma_1, \sigma_{1,l}$ then the result follows from 4.3.2. If $\rho_j = \sigma_1$ then ∂_j

126

$$\partial_j : N_n^G[\rho_j] \to N_{n-1}^G[F_{j-1}]$$

factors through

$$N_n^G[\rho_j] \cong N_{n-1}^G[\sigma_{-1}] \to N_{n-1}^G[F_{j-1}]$$

for the same reasons as given in section 4.2. Thus the result holds.

Finally, we have to consider the case $\rho_j = \sigma_{1,l}$. Look at the homomorphism

$$\partial_{1,l} : N_n^G[\sigma_{1,l}] \to N_{n-1}^G[F_{j-1}] \quad .$$

If $E \in N_n^G[\sigma_{1,l}]$ then $\partial_{1,l}(E)$ has slice types σ_{2l} and σ_{-1} ($\partial_{1,l}(E) = S(E)$) thus it makes sense to consider $\nu_{2l}\partial_{1,l}(E)$ which is an element of $N_{n-1}^G[\sigma_{2l}]$. There is therefore an N_* homomorphism

$$\nu_{2l}\partial_{1,l} : N_{n-1}^G[\sigma_{1,l}] \to N_{n-1}^G[\sigma_{2l}] \quad .$$

We also have the N_* homomorphism

$$q_{2l} : N_{n-1}^G[\sigma_{2l}] \to N_{n-1}^G[F_{j-1}]$$

$$E \to RP(R \times E) \quad .$$

Note that $q_{2l}(E)$ has slice types σ_{2l}, σ_1 and σ_{-1} so that $\nu_{2l}q_{2l}(E)$ makes sense and of course $\nu_{2l}q_{2l}(E) = E$. We assert that the following diagram

4.3.5
$$\begin{array}{ccc} N_n^G[\sigma_{1,l}] & \xrightarrow{\partial_{1,l}} & N_{n-1}^G[F_{j-1}] \\ {}_{\nu_{2l}\partial_{1,l}} \searrow & & \nearrow {}_{q_{2l}} \\ & N_{n-1}^G[\sigma_{2l}] & \end{array}$$

is commutative. The proof will use the fact that

$$\nu_{2l}(\partial_{1,l}(E)) = \nu_{2l}(q_{2l}(\nu_{2l}\partial_{1,l}(E)))$$

and the following lemma. In fact the next lemma proves more than we actually need.

4.3.6   LEMMA. *If M_1 and M_2 are n dimensional $\mathbb{Z}/4$ manifolds of type $\{\sigma_{2l}, \sigma_1, \sigma_{-1}\}$, if $\sigma_{2l}, \sigma_1, \sigma_{-1} \in F$ and if $\nu_{2l}(M_1)$ and $\nu_{2l}(M_2)$ are equivalent elements of $N_n^{\mathbb{Z}/4}[\sigma_{2l}]$ then M_1 and M_2 are F bordant.*

PROOF. Since $\nu_{2l}(M_1)$ and $\nu_{2l}(M_2)$ are bordant there is a bundle E over W of type σ_{2l} such that

$$E|\partial W = \nu_{2l}(M_1) + \nu_{2l}(M_2) \; .$$

Let N_1, N_2 be equivariant tubular neighbourhoods of the fixed point set $M_1^{\mathbb{Z}/2}$, $M_2^{\mathbb{Z}/2}$ respectively in M_1, M_2. We may identify $\partial(M_1 - N_1) + \partial(M_2 - N_2)$ with $\partial S(E)$ and so construct

$$M = (M_1 - N_1) + (M_2 - N_2) \cup S(E) \; .$$

This manifold M is a $\mathbb{Z}/4$ manifold of type $\{\sigma_1, \sigma_{-1}\}$ which is F bordant to $M_1 + M_2$. The bordism may be realized by

$$(M_1 + M_2) \times I \cup D(E)$$

where we glue $D(E)|\partial W$ to $(N_1 + N_2) \times \{1\}$. Finally the mapping cylinder $M \to M/(\mathbb{Z}/2)$ is a $\mathbb{Z}/4$ manifold with slice types σ_1, σ_{-1} and boundary precisely M. Hence M_1 and M_2 are F bordant.

We can now prove our assertion 4.3.5 quite easily. The manifolds $\partial_{1,l}(E)$ and $q_{2l}\nu_{2l}\partial_{1,l}(E)$ are of type $\{\sigma_{2l}, \sigma_1, \sigma_{-1}\}$ with

$$\nu_{2l}(\partial_{1,l}(E)) = \nu_{2l}(q_{2l}\nu_{2l}\partial_{1,l}(E))$$

and so by 4.3.6

$$\partial_{1,l}(E) \sim q_{2l}\nu_{2l}\partial_{1,l}(E) \text{ in } N^G_{n-1}[F_{j-1}].$$

This proves that the diagram in 4.3.5 is commutative.

Returning to the proof of 4.3.4, we have for $\rho_j = \sigma_{1,l}$:

$$N^G_n[F_j] \cong N^G_n[F_{j-1}]/(im\ \partial_{1,l}) \oplus ker\ \partial_{1,l}$$

since everything in question is a $\mathbb{Z}/2$ module. By assumption we have

$$N^G_n[\sigma_{0,l}] \oplus N^G_n[\sigma_{2l}] \oplus \ldots \cong N^G_n[F_{j-1}]$$

with the isomorphism, when restricted to $N^G_n[\sigma_{2l}]$ being given by q_{2l}. Thus

$$N^G_n[F_{j-1}]/(im\ \partial_{1,l}) \cong N^G_n[\sigma_{0,l}] \oplus N^G_n[\sigma_{2l}]/(im\ \nu_{2l}\partial_{1,l}) \oplus \ldots$$

because of 4.3.5. Furthermore, it is easy to see that $ker(\partial_{1,l}) \cong ker(\nu_{2l}\partial_{1,l})$ again because of 4.3.5. This completes the proof of Lemma 4.3.4.

To continue, we need to calculate the kernel and image of

$$\nu\partial : N^G_n[\sigma_{1,l}] \to N^G_{n-1}[\sigma_{2l}].$$

For $l = 0$ this is clearly an isomorphism so we need only consider $l \geq 1$. From our previous results we know that

$$N^G_*[\sigma_{1,l}] \cong N_{*-1-2l}(BO_1 \times BU_l)$$

and that N_* generators are given by

$$\xi_a \times \eta_{2J}$$

$$\xi_{a+2} \times \eta_{2b} \times \eta_I + \xi_{a+2} \times (\eta_{2b} \otimes \xi_{a+2}) \times \eta_I.$$

where $a, b \geq 0$, $2J = (2j_1, 2j_2, \ldots, 2j_l)$ with $j_1 \geq j_2 \geq \ldots \geq j_l$, $I = (i_2, i_3, \ldots, i_l)$ with $i_2 \geq i_3 \geq \ldots \geq i_l$ and $2b+1 \geq \max\{i_j; i_j \text{ odd}\}$. (See Lemma 3.4.19). Let K_* be the N_* submodule of $N_*^G[\sigma_{1,l}]$ generated by

$$\lambda(a, 2b+1, I) = \xi_{a+2} \times \eta_{2b} \times \eta_I + \xi_{a+2} \times (\eta_{2b} \otimes \xi_{a+2}) \times \eta_I.$$

Define a homomorphism

$$k: K_* \to N_*^G[F_j]$$

which on the basis elements is given by

$$\lambda(a, 2b+1, I) \to RP(R \times \xi_{a+2} \times \eta_{2b}) \times RP(R \times \eta_I).$$

A trivial calculation gives the next result.

4.3.7 LEMMA. *If $E \in K_*$ then $\nu k(E) = E$.*

4.3.8 COROLLARY. $K_* \subseteq Ker(\nu \partial) : N_n^G[\sigma_{1,l}] \to N_{n-1}^G[\sigma_{2l}]$.

In fact we can say more.

4.3.9 LEMMA. $K_* = Ker(\nu \partial)$.

To prove this recall that the group $\Gamma(\sigma_{2l})$ is precisely $\mathbb{Z}/4 \times_{\mathbb{Z}/2} O_{2l}$ and hence N_* generators of $N_*^G[\sigma_{2l}]$ are

$$\mathbb{Z}/4 \times_{\mathbb{Z}/2} \xi_I, \quad S^1 \times_{\mathbb{Z}/2} \xi_I, \quad S^a \times \eta_{2J}$$

where $I = (i_1, i_2, \ldots, i_l)$, $2J = (2j_1, 2j_2, \ldots, 2j_l)$.

The homomorphism

$$\nu\partial : N^G_*[\sigma_{1,l}] \to N^G_*[\sigma_{2l}]$$

is given by

$$E \to S(E) \to \nu S(E).$$

where ν denotes taking the normal bundle of the points with slice type σ_{2l}. Now, if $\xi_a \times \eta_{2J} \in N^G_*[\sigma_{1,l}]$

$$\nu\partial(\xi_a \times \eta_J) = S^a \times \eta_{2J}.$$

Thus the image is precisely generated by $S^a \times \eta_{2J}$ and the kernel is precisely K_*.

We can now prove Theorem 4.3.1 as follows. The following homomorphisms

(i) $\quad N^G_n[\sigma_k] \to N^G_n \qquad , k \neq 1, k$ odd

$\qquad\qquad E \to RP(R \times E)$

(ii) $\quad N^G_n[\sigma_{k,l}] \to N^G_n \qquad , k \neq 1$

$\qquad\qquad E \to RP(R \times E)$

(iii) $\quad N^G_n[\sigma_{2l}]/im(\nu\partial) \to N^G_n$

$\qquad\qquad E \to RP(R \times E)$

(iv) $\quad ker(\nu\partial) \to N^G_n$

$\qquad\qquad E \to k(E)$

together realise the isomorphism of Lemma 4.3.3. Theorem 4.3.1 then follows.

4.3.10 REMARK. By using slightly different homomorphisms in (i)-(iv) above we can replace

$RP(R\times\xi_I\times\eta_J)$ $(I = (i_1, i_2, \ldots, i_k)$ with $k \neq 1)$

by

$RP(R\times\xi_I) \times RP(R\times\eta_J)$

and replace

$RP(R\times\eta_J)$

by

$RP(R\times\eta_{j(1)}) \times RP(R\times\eta_{j(2)}) \times \ldots \times RP(R\times\eta_{j(l)})$.

4.3.11 REMARK. There is another splitting homomorphism

$$q_{2l} : N_*^G[\sigma_{2l}] \to N_*^G[F_j]$$

which is defined on the basis elements of $N_*^G[\sigma_{2l}]$ by

$\mathbb{Z}/4 \times_{\mathbb{Z}/2} \xi_I \longmapsto \mathbb{Z}/4 \times_{\mathbb{Z}/2} RP(R\times\xi_I)$

$S^1 \times_{\mathbb{Z}/2} \xi_I \longmapsto S^1 \times_{\mathbb{Z}/2} RP(R\times\xi_I)$

$S^a \times \eta_{2J} \longmapsto S(\xi_a \times \eta_{2J})$.

This splitting dispenses (in some sense) the calculations in 4.3.5 and 4.3.6.

4.4 GENERATORS FOR $\mathbb{Z}/2^r$ BORDISM

In this section we shall find the generators of the $\mathbb{Z}/2^r$ bordism ring for $r > 2$. The method is a straightforward generalisation of that in section 4.3 for the case of $\mathbb{Z}/4$ bordism.

4.4.1 THEOREM. $N_*^{\mathbb{Z}/2^r}$ is a free N_* module which is generated as an N_* module by the following $\mathbb{Z}/2^r$ manifolds:

$$X \times_{\mathbb{Z}/2^s} RP(R \times (\xi_I \otimes \tilde{R})) \times \prod_{i=1}^{2^{s-1}-1} RP(R \times (\eta_{J(i)} \otimes V_i)) \text{ for } 1 \leq s \leq r.$$

$$X \times_{\mathbb{Z}/2^s} RP(R \times (\xi_{2a} \otimes \tilde{R}) \times (\eta_{J(c)} \otimes V_c)) \times \prod_{i=1}^{c-1} RP(R \times (\eta_{J(i)} \otimes V_i))$$

$$\text{for } a \geq 0, \; 2^{s-2} < c \leq 2^{s-1}-1, \; 2 \leq s \leq r.$$

$$X \times_{\mathbb{Z}/2^s} RP(R \times (\xi_{2a+2} \otimes \tilde{R}) \times (\eta_{2b} \otimes V_c)) \times \prod_{i=1}^{c} RP(R \times (\eta_{J(i)} \otimes V_i))$$

$$\text{for } a,b \geq 0, \; c = 2^{s-2}, \; 2 \leq s \leq r,$$

where if $r \neq s$, then $X = \mathbb{Z}/2^r$ or S^1 and otherwise $X = \mathbb{Z}/2^r$; $I = (i_1, i_2, \ldots, i_k)$ with $i_1 \geq i_2 \geq \ldots \geq i_k \geq 0$, $k \geq 0$, $k \neq 1$ and if $s \neq r$ then $k \geq 2$; $J(i) = (j(i,1), j(i,2), \ldots, j(i,l_i))$ with $j(i,1) \geq j(i,2) \geq \ldots \geq j(i,l_i) \geq 0$ and $l_i \geq 0$.

In the above V_j ($j = 1, 2, \ldots$) denotes the $\mathbb{Z}/2^r$ module with underlying space the complex numbers and with action of the generator of $\mathbb{Z}/2^r$ given by multiplication by $\exp(2\pi i j/2^r)$. The group $\mathbb{Z}/2^r$ and its subgroups $\mathbb{Z}/2^s$ act on X in the obvious way. The subgroup $\mathbb{Z}/2^s$ acts on $\tilde{R}$ (underlying space R) by multiplication by -1 and acts on V_c (underlying space C) by the restriction of the $\mathbb{Z}/2^r$ action, i.e., multiplication by $\exp(2\pi i c/2^s)$.

For the proof of the theorem we shall need to know the $\mathbb{Z}/2^r$-slice types. Note that the non-trivial irreducible $\mathbb{Z}/2^s$ modules are

$$V_0, V_1, \ldots, V_{2^{s-1}-1}$$

where V_0 denotes the real numbers with generator of $\mathbb{Z}/2^s$ acting by multiplication by -1 (i.e., $V_0 = \tilde{R}$ in the notation of 4.4.1), while V_j $(j \geq 1)$ denotes the complex numbers with the generator of $\mathbb{Z}/2^s$ acting by multiplication by $\exp(2\pi i j / 2^s)$. The $\mathbb{Z}/2^r$-slice types are therefore of the form

$$\sigma_A^s = \sigma^s(a(0), a(1), \ldots, a(2^{s-1}-1))$$
$$= \left[\mathbb{Z}/2^s; \prod_j V_j^{a(j)}\right]$$

where $0 \leq s \leq r$ and A is a 2^{s-1} tuple of non-negative integers. We put a total ordering on the set of $\mathbb{Z}/2^r$-slice types by the method described in section 1.7. In fact we just need any ordering which satisfies the following three conditions:

(i) $\sigma^s(a_0, a_1, a_2, \ldots) > \sigma^s(1, a_1, a_2, \ldots)$ if $a_0 > 1$

(ii) $\sigma^s(1, a(1), \ldots, a(j), \ldots, a(2^{s-1}-j), \ldots) >$
$\sigma^s(1, a(1), \ldots, a(j-1), a(j)+a(2^{s-1}-j), a(j+1), \ldots$
$\ldots, a(2^{s-1}-j-1), 0, a(2^{s-1}-j+1), \ldots)$
if $a(2^{s-1}-j) > 0$ and $j < 2^{s-2}$.

(iii) $\sigma^{s+1}(a(0), a(1), \ldots, a(2^s-1)) >$
$\sigma^s(2a(2^{s-1}), a(1)+a(2^s-1), a(2)+a(2^s-2), \ldots)$.

We now rename the $\mathbb{Z}/2^r$-slice types:

$$\rho_0, \rho_1, \rho_2, \ldots$$

with the condition that if $i < j$ then $\rho_i < \rho_j$. Let F_j be defined by

$$F_j = \{\rho_0, \rho_1, \ldots, \rho_j\} \quad .$$

It is not difficult to verify that F_j is a family of $\mathbb{Z}/2^r$-slice types.

4.4.2 LEMMA. *If $\rho_j = \sigma_A^s$ with $A = (a(0), a(1), \ldots)$ satisfying either $a(0) \neq 1$ or $a(0) = 1$ and $a(i) \neq 0$ for some $i > 2^{s-2}$ then*

$$0 \to N_n^G[F_{j-1}] \to N_n^G[F_j] \to N_n^G[\rho_j] \to 0$$

is a short split exact sequence, where $G = \mathbb{Z}/2^r$.

PROOF. In the case that $a(0) \neq 1$ this follows from Corollary 1.7.3 with a splitting homomorphism given by

$$N_n^G[\rho_j] \longrightarrow N_n^G[F_j]$$

$$E \longmapsto RP(R \times E) \quad .$$

There is another splitting homomorphism. To define it recall that a G vector bundle over a trivial G space splits into a direct sum of vector bundles according to the action of G -- see [Atiyah; Proposition 1.6.2] or [Conner and Floyd, 1]. If $E \in N_*^G[\sigma_A^s]$ then write E as

$$E = E_0 \oplus E_1$$

where E_0 corresponds to the subbundle with $\mathbb{Z}/2^s$ action in the fibre determined by $V_0^{a(0)}$ and E_1 corresponds to $\prod_{j \geq 1} V_j^{a(j)}$. If the base space of E is N then $RP(R \times E_0)$ is an $RP(R \times V_0^{a(0)})$ bundle over N. We have natural maps

135

$$\begin{array}{c} \phantom{RP(R\times E_0)\xrightarrow{f}} E_1 \\ \phantom{RP(R\times E_0)\xrightarrow{f}} \downarrow \\ RP(R\times E_0) \xrightarrow{f} N \end{array}$$

Then we define $q(E)$ to be $RP(R\times f^*(E_1))$. It is convenient to use the notation $RP(R\times(E_0,E_1))$. Note that $q(E)$ is an

$$RP(R\times V_0^{a(0)}) \times RP(R\times \prod_{0\geq 1} V_j^{a(j)})$$

bundle over N. For example, note that

$$q(\mathbb{Z}/2^r \times_{\mathbb{Z}/2^s} (\xi_I \otimes V_0) \times \prod_{i=1}^{2^{s-1}-1} (\eta_{J(i)} \otimes V_i))$$

$$= \mathbb{Z}/2^r \times_{\mathbb{Z}/2^s} RP(R\times(\xi_I \otimes V_0)) \times RP(R\times \prod_{i=1}^{2^{s-1}-1} (\eta_{J(i)} \otimes V_i)).$$

We leave it for the reader to check that this definition of q does indeed give a splitting.

If $a(0) = 1$ and $a(j) \neq 0$ for some $j > 2^{s-2}$ then a splitting may be obtained by writing $E \in N_*^G[\sigma_A^s]$ as

$$E = E_{0,c} \oplus E_1$$

where $E_{0,c}$ is the subbundle with action determined by $V_0 \times V_c^{a(c)}$ where $c = max\{j; a_j \neq 0\}$. We define $q(E)$ by the method just described, i.e.,

$$q(E) = RP(R\times(E_{0,c}, E_1))$$

so that $q(E)$ is an

$$RP(R\times V_0 \times V_c^{a(c)}) \times RP(R\times \prod_{j=1}^{c-1} V_j^{a(j)})$$

bundle over some manifold. The second condition (ii) for the ordering of $\mathbb{Z}/2^r$-slice types ensures that this definition of q provides a splitting, again all details are left for the reader. (The main point being that $V_0 \otimes V_c = V_d$ where $d = 2^{s-1} - c$.)

We shall shortly state a result analogous to Lemma 4.3.4, but first we need to define subsets A_j, B_j of F_j. We define these inductively as follows:

$$A_0 = \{\rho_0\} \quad , \quad B_0 = \emptyset \;.$$

Suppose that A_{j-1}, B_{j-1} are defined. There are two possibilities for ρ_j:

(i) $\rho_j = \sigma_A^s$ with $A = (a(0), a(1), \ldots)$ and either $a(0) \neq 1$ or $s > 1$, $a(0) = 1$ and $a(i) \neq 0$ for some $i > 2^{s-2}$.

(ii) $\rho_j = \sigma_A^s$ with A taking the form
$$(1, a(1), a(2), \ldots, a(2^{s-2}), 0, 0, \ldots, 0)$$
if $s \geq 2$, and $A = (1)$ if $s = 1$.

In case (i) we define

$$A_j = A_{j-1} \cup \{\rho_j\}$$
$$B_j = B_{j-1}$$

while in case (ii) we define

$$A_j = A_{j-1} - \{\sigma_B^{s-1}\}$$
$$B_j = B_{j-1} \cup \{\sigma_A^s\}$$

where B is given by

$$B = (2a(2^{s-2}), a(1), a(2), \ldots, a(2^{s-2}-1))$$

if $s \geq 2$ and $B = (0)$ if $s = 1$.

Also, set $r(\sigma_A^s) = \sigma_B^{s-1}$. These definitions make sense and

$$A_j \cup B_j \cup \{r(\rho); \rho \in B_j\} = F_j .$$

If $\sigma_B^{s-1} = r(\sigma_A^s)$ then there is an N_* homomorphism

$$N_n^G[\sigma_A^s] \xrightarrow{\nu\partial} N_{n-1}^G[\sigma_B^{s-1}]$$

where $\nu\partial(E) = \nu(S(E))$.

4.4.3 THEOREM. *If $G = \mathbb{Z}/2^r$ then $N_n^G[F_j]$ is isomorphic to the direct sum of the following*

(i) $\bigoplus\limits_{\rho \in A_j} N_n^G[\rho]$

(ii) $\bigoplus\limits_{\rho \in B_j} (ker(\nu\partial : N_n^G[\rho] \to N_{n-1}^G[r(\rho)]))$

(iii) $\bigoplus\limits_{\rho \in B_j} N_n^G[r(\rho)]/(im(\nu\partial : N_{n+1}^G[\rho] \to N_n^G[r(\rho)]))$.

Furthermore, the isomorphism from the direct sum to $N_n^G[F_j]$, when restricted to the terms in (i) and (iii) is given by the splitting homomorphism of **4.4.2**.

PROOF. The proof is by induction. The result is clear for $j = 0$. Suppose that the result holds for $j-1$ ($j \geq 1$) then we have the exact sequence

$$\ldots \to N_n^G[F_{j-1}] \to N_n^G[F_j] \xrightarrow{\nu_j} N_n^G[\rho_j] \xrightarrow{\partial_j} N_{n-1}^G[F_{j-1}] \to \ldots .$$

If $\rho_j \in A_j$ then the result follows from 4.4.2. If $\rho_j \in B_j$ then in analogy to 4.3.5 we claim that the following diagram

4.4.4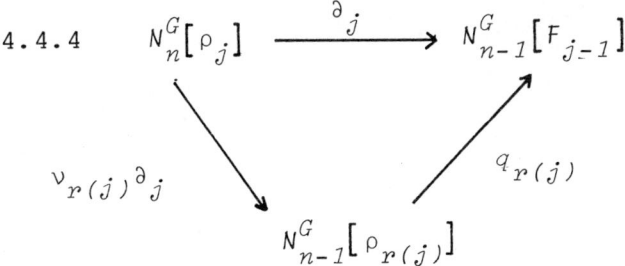

is commutative, where $r(j)$ is given by $\rho_{r(j)} = r(\rho_j)$ and $q_{r(j)}(F) = RP(R \times F)$ or $RP(R \times (F_0, F_1))$. Observe that

$$\nu_{r(j)} \partial_j : N_n^G[\rho_j] \to N_{n-1}^G[\rho_{r(j)}]$$

is well defined. To prove 4.4.4 we consider an element $E \in N_n^G[\rho_j]$. Write ρ_j as

$$\rho_j = [\mathbb{Z}/2^s; V_0 \times V]$$

where $V_0 \not\subseteq V$. The bundle E splits as

$$E = E_0 \oplus E_1$$

corresponding to the decomposition of ρ_j. Let $\mathbb{Z}/2$ act on E by multiplication by -1 in each fibre of E_1. This defines an involution on $\partial_j(E) = S(E)$. Similiarly we get an involution on $\nu_{r(j)} \partial_j(E)$ and on $q_{r(j)} \nu_{r(j)} \partial_j(E)$. Since

$$\nu_{r(j)} \partial_j(E) = \nu_{r(j)} q_{r(j)} \nu_{r(j)} \partial_j(E)$$

(because $\nu_{r(j)} q_{r(j)} = 1$) we can construct a $\mathbb{Z}/2^r$ manifold M by glueing together

$$\partial_j(E) - D(\nu_{r(j)} \partial_j(E))$$

and

$$q_{r(j)} \nu_{r(j)} \partial_j(E) - D(\nu_{r(j)} \partial_j(E)).$$

Furthermore M comes equipped with an involution. The manifold $\partial_j(E)$ is in fact a fibre bundle with fibre

$$\mathbb{Z}/2^r \times_{\mathbb{Z}/2^s} S(V_0 \times V)$$

while $q_{r(j)} \nu_{r(j)} \partial_j(E)$ is a fibre bundle with fibre

$$\mathbb{Z}/2^r \times_{\mathbb{Z}/2^s} RP(R \times V).$$

The involution in each case is multiplication by -1 in V. It is then easy to see that the involution on M has only a codimension one fixed point set. Therefore the mapping cylinder N of

$$M \to M/(\text{Involution})$$

is a $\mathbb{Z}/2^r$ manifold with boundary M. A straightforward calculation shows that the isotropy subgroups in N are contained in $\mathbb{Z}/2^s$ and so N is a $\mathbb{Z}/2^r$ manifold of type F_{j-1}. This proves 4.4.4 and Theorem 4.4.3 follows immediately.

To prove Theorem 4.4.1 we argue as in section 4.3. From chapter 3 we have the following

4.4.5 LEMMA. *Generators of* $N_*^G[\sigma_A^s]$ *as an* N_* *module are given by the following*

$$\mathbb{Z}/2^r \times_{\mathbb{Z}/2^s} (\xi_I \otimes V_0) \times \prod_{i=1}^{2^{s-1}-1} (\eta_{J(i)} \otimes V_i)$$

$$S^1 \times_{\mathbb{Z}/2^s} (\xi_I \otimes V_0) \times \Pi(\eta_{J(i)} \otimes V_i)$$

$$\mathbb{Z}/2^r \times_{\mathbb{Z}/2^{s+1}} S^{2i} \times (\eta_{2K} \otimes V(2^{s-1})) \times \Pi(\eta_{J(i)} \otimes V_i)$$

$$S^1 \times_{\mathbb{Z}/2^{s+1}} S^{2i} \times (\eta_{2K} \otimes V(2^{s-1})) \times \Pi(\eta_{J(i)} \otimes V_i)$$

where the last two types occur only if $r > s$ and $a(0)$ is even; $S^{2i} = S(V_0^{2i+1})$. The symbols $I, K, J(i)$ respectively are non-increasing sequences of non-negative integers of length $a(0)$, $a(0)/2$, $a(i)$ respectively. Furthermore, if $a(0) = 1$ then

$$\mathbb{Z}/2^r \times_{\mathbb{Z}/2^s} (\xi_{2a} \otimes V_0) \times \Pi(\eta_{J(i)} \otimes V_i)$$

$$S^1 \times_{\mathbb{Z}/2^s} (\xi_{2a} \otimes V_0) \times \Pi(\eta_{J(i)} \otimes V_i)$$

for $a \geq 0$ give an N_* basis.

We need to calculate the kernel and image of the homomorphisms

$$\nu\partial : N_n^G[\rho] \to N_{n-1}^G[r(\rho)]$$

for $\rho \in B_j$. Let $\rho = \sigma_A^s \in B_j$ where

$$A = (1, a(1), a(2), \ldots, a(2^{s-2}), 0, 0, \ldots, 0).$$

Two cases arise:

(1) $s = 1$ or $s \geq 2$ and $a(2^{s-2}) = 0$

(2) $s \geq 2$ and $a(2^{s-2}) \neq 0$.

In case (1) an easy calculation shows that

$$\nu\partial : N_n^G[\rho] \to N_{n-1}^G[r(\rho)]$$

is an isomorphism. (For example, observe that

$$N_n^G[\rho] \cong N_{n-1-\Sigma a(i)}(B(\mathbb{Z}/2^r \times_{\mathbb{Z}/2^s} O_1) \times \prod_{i=1}^{2^{s-2}-1} BU_{a(i)})$$

and

$$N_{n-1}^G[r(\rho)] \cong N_{n-1-\Sigma a(i)}(B((\mathbb{Z}/2^r)/(\mathbb{Z}/2^{s-1}))) \times \prod_{i=1}^{2^{s-2}-1} BU_{a(i)})$$

the latter being true because $r(\rho) = \sigma_B^{s-1}$ where

141

$$B = (0, a(1), a(2), \ldots, a(2^{s-2}-1)).$$

Then check what happens to the isomorphisms under $\nu\partial$.)

In case (2) let $K_*[\sigma_A^s]$ be the N_* submodule of $N_*^G[\sigma_A^s]$ generated by

$$\lambda(X, 2a, 2b+1, I, J(1), J(2), \ldots, J(2^{s-2}-1)) =$$

$$X \times_{\mathbb{Z}/2^s} (\xi_{2a+2} \otimes V_0) \times (\eta_{2b} \otimes V(2^{s-2})) \times (\eta_I \otimes V(2^{s-2})) \times \prod_{i=1}^{2^{s-2}-1} (\eta_{J(i)} \otimes V_i)$$

$$+ X \times_{\mathbb{Z}/2^s} (\xi_{2a+2} \otimes V_0) \times ((\eta_{2b} \otimes V(2^{s-2})) \times (\xi_{2a+2} \otimes V_0)) \times (\eta_I \otimes V(2^{s-2}))$$

$$\times \prod_{i=1}^{2^{s-2}-1} (\eta_{J(i)} \otimes V_i)$$

where $X = \mathbb{Z}/2^r$ or S^1, $a, b \geq 0$, $I = (i_2, i_3, \ldots, i_l)$, $(l = a(2^{s-2}))$, with $i_2 \geq i_3 \geq \ldots \geq i_l$ and $2b+1 \geq \max\{i_j; i_j \text{ odd}\}$. Note that

$$X \times_{\mathbb{Z}/2^s} (\xi_{2a} \otimes V_0) \times (\eta_{2J} \otimes V(2^{s-2})) \otimes \prod_{i=1}^{2^{s-2}-1} (\eta_{J(i)} \otimes V_i)$$

and

$$\lambda(X, 2a, 2b+1, I, J(1), J(2), \ldots, J(2^{s-2}-1))$$

form an N_* basis for $N_*^G[\sigma_A^s]$, this follows from 4.4.5 and 3.4.19. We can then show that

$$K_*[\sigma_A^s] = \ker(\nu_{r(j)} \partial_j),$$

in fact the splitting $k: K_*[\sigma_A^s] \to N_*^G[F_j]$ is given by

$$\lambda(X, 2a, 2b+1, I, J(1), \ldots, J(2^{s-2}-1)) \longmapsto$$

$$X \times_{\mathbb{Z}/2^s} RP(R \times (\xi_{2a+2} \otimes V_0) \times (\eta_{2b} \otimes V(2^{s-2}))) \times$$
$$RP(R \times (\eta_I \otimes V(2^{s-2}))) \times \prod_{i=1}^{2^{s-2}-1} RP(R \times (\eta_{J(i)} \otimes V_i)) \quad .$$

If we let $L_*[\sigma_A^s]$ be the N_* submodule of $N_*^G[\sigma_A^s]$ generated by

$$X \times_{\mathbb{Z}/2^s} (\xi_{2a} \times V_0) \times (\eta_{2J} \otimes V(2^{s-2})) \times \prod_{i=1}^{2^{s-2}-1} (\eta_{J(i)} \otimes V_i)$$

then

$$N_*^G[\sigma_A^s] \cong K_*[\sigma_A^s] \oplus L_*[\sigma_A^s]$$

and $L_*[\sigma_A^s]$ maps monomorphically under $\nu_{r(j)} \partial_j$ into $N_*^G[r(\sigma_A^s)]$. The remainder of the proof of Theorem 4.4.1 is the same as that of 4.3.1 in section 4.3.

4.4.6 REMARK. As in 4.3.11 we could use a different splitting for $N_*^G[r(\rho)]$, $\rho \in B_j$. This is given on the generators by

$$X \times_{\mathbb{Z}/2^s} (\xi_I \otimes V_0) \times \Pi(\eta_{J(i)} \otimes V_i) \longmapsto$$

$$X \times_{\mathbb{Z}/2^s} RP(R \times (\xi_I \otimes V_0)) \times RP(R \times (\eta_{J(i)} \otimes V_i))$$

$$X \times_{\mathbb{Z}/2^{s+1}} S^{2i} \times (\eta_{2k} \otimes V(2^{s-1})) \times \Pi(\eta_{J(i)} \otimes V_i) \longmapsto$$

$$X \times_{\mathbb{Z}/2^{s+1}} S((\xi_{2i} \otimes V_0) \times (\eta_{2K} \otimes V(2^{s-1})) \times \Pi(\eta_{J(i)} \otimes V_i)) \quad .$$

4.4.7 REMARK. In the main theorem we used the fact that

$$RP(R \times \prod_i (\eta_{J(i)} \otimes V_i))$$

could be replaced by

$$\prod_i RP(R \times (\eta_{J(i)} \otimes V_i)).$$

We could make other changes. We leave the following result for the reader to prove.

4.4.8 EXERCISE. Prove that generators of N_*^G, $G = \mathbb{Z}/2^r$, as an N_* module may be obtained from products of the following $\mathbb{Z}/2^r$ manifolds

$$X \times_{\mathbb{Z}/2^s} RP(R \times (\xi_I \otimes \tilde{R})) \quad \text{for } 1 \leq s \leq r$$

$$X \times_{\mathbb{Z}/2^s} RP(R \times (\xi_{2a} \otimes \tilde{R}) \times (\eta_b \otimes V_c))$$

for $2 \leq s \leq r$, $2^{s-2} \leq c \leq 2^{s-1}$, $a \geq 0$ furthermore if $c = 2^{s-2}$ then $a > 0$ and b is even,

$$RP(R \times (\eta_j \otimes V_i)) \text{ for } j \geq 0 \text{ and } 1 \leq i \leq 2^{r-1}-1,$$

where $X = \mathbb{Z}/2^r$ or S^1 if $s \neq r$ and $X = \mathbb{Z}/2^r$ if $s = r$.

In general our goal has been to write down generators of N_*^G. It is not difficult to write down generators of $N_*^G[F_j]$ where the F_j are as defined at the beginning of this section. Furthermore if F is a family of subgroups of G (see 1.3.1) then it is an easy calculation to find generators of $N_*^G(F)$, we leave this as an exercise for the reader. For example:

4.4.9 EXERCISE. Suppose that $G = \mathbb{Z}/2^r$ and that $F_q = \{\mathbb{Z}/2^s; 0 \leq s \leq q \leq r\}$. Prove that $N_*^G(F_q)$ is a free N_* module which is generated by

$$X \times_{\mathbb{Z}/2^s} RP(R \times (\xi_I \otimes \tilde{R})) \times \prod_{i=1}^{2^{s-1}-1} RP(R \times (\eta_{J(i)} \otimes V_i))$$

for $1 \leq s \leq q$

$$X \times_{\mathbb{Z}/2^s} RP(R \times (\xi_{2a} \otimes \tilde{R}) \times (\eta_{J(i)} \otimes V_i)) \times \prod_{i=1}^{c-1} RP(R \times (\eta_{J(i)} \otimes V_i))$$

for $a \geq 0$, $2^{s-2} < c \leq 2^{s-1}-1$, $2 \leq s \leq q$.

$$X \times_{\mathbb{Z}/2^s} RP(R \times (\xi_{2a+2} \otimes \tilde{R}) \times (\eta_{2b} \otimes V_c)) \times \prod_{i=1}^{c} RP(R \times (\eta_{J(i)} \otimes V_i))$$

for $a, b \geq 0$, $c = 2^{s-2}$, $2 \leq s \leq q$

and if $q < r$ then also

$$X \times_{\mathbb{Z}/2^{q+1}} S^{2a} \times RP(R \times (\eta_{2K} \otimes V_d)) \times \prod_{i=1}^{c-1} RP(R \times (\eta_{J(i)} \otimes V_i))$$

for $a \geq 0$, $d = 2^{q-1}$

where $X = \mathbb{Z}/2^r$ or S^1 (unless $s = r$ in which case $X = \mathbb{Z}/2^r$); $I = (i_1, i_2, \ldots, i_k)$ with $i_1 \geq i_2 \geq \ldots \geq i_k \geq 0$, $k \geq 0$, $k \neq 1$ and if $s \neq q$ then $k \geq 2$; $J(i) = (j(i,1), j(i,2), \ldots, j(i,l_i))$ with $j(i,1) \geq j(i,2) \geq \ldots \geq j(i,l_i) \geq 0$ and $l_i \geq 0$.

Alternatively the last subset of generators may be replaced by the manifolds

$$X \times_{\mathbb{Z}/2^{q+1}} S((\xi_{2a} \otimes \tilde{R}) \times (\eta_{2K} \otimes V_d) \times \prod_{i=1}^{c-1} (\eta_{J(i)} \otimes V_i))$$

or some such variant.

4.5 $(\mathbb{Z}/2)^k$ BORDISM - ALGEBRAIC RESULTS

One of the main objects of this section will be to establish the following result.

4.5.1 THEOREM. *If* $G = (\mathbb{Z}/2)^k$ *then*

$$N_n^G \cong \bigoplus_{J \in S^n} N_{n-|J|}(BO_J)$$

where $J = (j(1), j(2), \ldots, j(2^k-1))$ *is a* (2^k-1) *tuple of non-negative integers,* $|J| = \Sigma j(i)$ *and* $BO_J = \Pi BO_{j(i)}$.

The set S^n is defined next.

4.5.2 DEFINITION. The set S^n is the subset of

$$\{J = (j(1), j(2), \ldots, j(2^k-1); j(i) \geq 0 \text{ and } |J| \leq n\}$$

consisting of those (2^k-1) tuples $(j(1), j(2), \ldots)$ which satisfy the condition:

For all integers a such that

$$\sum_{i=2^a}^{2^{a+1}-1} j(i) = 1$$

say $j(m) = 1$, $(2^a \leq m \leq 2^{a+1})$, there is an integer $l \neq m$, $(1 \leq l \leq 2^k-1)$ such that $j(l) \neq 0$ and $m*l < l$.

Here $m*l$ is defined in the following way: Write m and l in their "2-adic form"

$$m = m_1 2^{k-1} + m_2 2^{k-2} + \ldots + m_{k-1} 2 + m_k$$
$$l = l_1 2^{k-1} + l_2 2^{k-2} + \ldots + l_{k-1} 2 + l_k .$$

Although this is not the usual way of taking the 2-adic forms of integers it is convenient for our purpose. It does make sense because k will be fixed and we shall only take 2-adic forms of integers less than 2^k. The integer $m*l$ is defined to be the unique integer with 2-adic form

$$p_1 2^{k-1} + p_2 2^{k-2} + \ldots + p_{k-1} 2 + p_k$$

where p_i is the mod 2 reduction of $m_i + l_i$, in other words

$$p_i = \tfrac{1}{2}(1-(-1)^{m_i+l_i}).$$

For example, if $k = 2$, i.e., $G = (\mathbb{Z}/2)^2$ then the set S^n is precisely

$$\{(a,b,c); a \neq 1, \ b+c \neq 1, \ a+b+c \leq n\} \cup$$
$$\{(1,b,c); c \neq 0, \ b+c \neq 1, \ 1+b+c \leq n\}.$$

4.5.3 REMARK. There is a one-to-one correspondence between the integers $0, 1, \ldots, 2^k - 1$ and elements of $G = (\mathbb{Z}/2)^k$ via the 2-adic expansion of integers less than 2^k

if $m = m_1 2^{k-1} + m_2 2^{k-2} + \ldots + m_k$

then $x_m = \sum_{i=1}^{k} m_i g_i$

where $g_1, g_2, \ldots, g_k$ is a preferred basis for $G = (\mathbb{Z}/2)^k$. Under this correspondence $m*n$ corresponds to $x_m + x_n$.

The proof of Theorem 4.5.1 will occupy the rest of this section. Throughout, G will denote $(\mathbb{Z}/2)^k$. The idea of the proof is roughly as follows. We index the G slice types in some natural way and define families F_j, $j \geq 0$ of G slice types so that

$$F_0 \subseteq F_1 \subseteq F_2 \subseteq \ldots$$

and $F_j - F_{j-1}$ consists of a single slice type. For each family F_j we define subsets A_j, B_j and a function

$$r : B_j \to F_j$$

such that

$$A_j \cup B_j \cup r(B_j) = F_j.$$

Then we prove that $N_n^G[F_j]$ is the direct sum of the following

$$\bigoplus_{\rho \in A_j} N_n^G[\rho]$$

$$\bigoplus_{\rho \in B_j} \ker(\nu \partial : N_n^G[\rho] \to N_{n-1}[r(\rho)])$$

$$\bigoplus_{\rho \in B_j} N_n^G[r(\rho)] / (im(\nu \partial : N_{n+1}^G[\rho] \to N_n^G[r(\rho)])) \quad .$$

It turns out that $\nu \partial$ is an isomorphism in each case and so

$$N_n^G[F_j] \cong \bigoplus_{\rho \in A_j} N_n^G[\rho].$$

Finally we obtain the main theorem after analysing the set A_j.

In order to index the G slice types we shall need to index the irreducible H modules for all $H \subseteq G$ and then index all the subgroups of G in some natural way. We start by looking at the irreducible G modules. If H is a subgroup of G with G/H cyclic then H is of index 2 in G. The converse is also true, trivially. We see from section 1.6 that the non-trivial irreducible G modules are precisely

$$\{V_H; H \subseteq_2 G\}$$

where $\subseteq_2$ means a subgroup of index 2 and V_H ($=V(H,1)$ in previous notation) is the real numbers with action of $g \in G$ given by multiplication by

$$\begin{cases} +1 & g \in H \\ -1 & g \notin H \end{cases}.$$

Notice that there are precisely 2^k-1 non-trivial irreducible G modules. We shall presently index, in a natural way, the non-trivial irreducible G modules by the integers $1, 2, \ldots, 2^k-1$. Let $g_1, g_2, \ldots, g_k$ be a basis for G and let H be a subgroup of G index 2. If

$$g_{i(1)}, g_{i(2)}, \ldots, g_{i(l)} \in H$$

for $0 \leq l < k$, $i(1) < i(2) < \ldots < i(l)$ and if

$$g_{j(1)}, g_{j(2)}, \ldots, g_{j(k-l)} \notin H$$

for $j(1) < j(2) < \ldots < j(k-l)$ then let

$$n(H) = \sum_{i=1}^{k-l} 2^{k-j(i)} .$$

The integer $n(H)$ satisfies $0 < n < 2^k$. Conversely, given an integer n (less than 2^k), write n in its 2-adic form

$$n = n_1 2^{k-1} + n_2 2^{k-2} + \ldots + n_{k-1} 2 + n_k$$

with $n_j = 0$ or 1. Define $H(n)$ to be the subgroup with basis

$$\{g_j ; n_j = 0\} \cup \{g_m + g_l ; m = \min_j \{j ; n_j = 1\}, n_l = 1, l > m\}.$$

The fact that $n(H(n)) = n$ is immediate while the fact that $H(n(H)) = H$ follows from the next lemma.

4.5.4 LEMMA. *Let H be a subgroup of $G = (\mathbb{Z}/2)^k$ of index 2. If*

$$g_{i(1)}, g_{i(2)}, \ldots, g_{i(l)} \in H$$

for $0 \leq l < k$, $i(1) < i(2) < \ldots < i(l)$ and if

$$g_{j(1)}, g_{j(2)}, \ldots, g_{j(k-l)} \notin H$$

for $j(1) < j(2) < \ldots < j(k-l)$ then

$$\{g_{i(j)}; 1 \leq j \leq l\} \cup \{g_{j(1)} + g_{j(i)}; 1 < i \leq k-l\}$$

is a basis for H.

PROOF. Consider $(\mathbb{Z}/2)^k$ as a $\mathbb{Z}/2$ vector space with basis $g_1, g_2, \ldots, g_k$, then subgroups of index 2 correspond to subspaces of rank $k-1$. Define an inner product $(\ ,\ )$ on G by

$$(\Sigma \alpha_i g_i, \Sigma \beta_i g_i) = \Sigma \alpha_i \beta_i$$

where $\alpha_i \in \mathbb{Z}/2 = \{0,1\}$. The conditions on H tell us that H is the subspace orthogonal to the vector $g = g_{j(1)} + g_{j(2)} + \ldots + g_{j(k-l)}$. But the subspace with basis

$$\{g_{i(j)}; 1 \leq j \leq l\} \cup \{g_{j(1)} + g_{j(i)}, 1 < i \leq k-l\}$$

is orthogonal to the vector g and hence coincides with H.

Out of interest, we can deduce the following result by induction.

4.5.5. COROLLARY. *Let K be a subgroup of G of index 2^l then there is a basis of K with elements of the form*

$$g_{j(1)} + g_{j(2)} + \ldots + g_{j(m)}, \quad m \leq l+1 .$$

4.5.6 LEMMA. *The non-trivial irreducible $G = (\mathbb{Z}/2)^k$ modules are*

$$\{V_n; 1 \leq n \leq 2^k - 1\}$$

where V_n has underlying space R and if

$$n = n(1)2^{k-1} + n(2)2^{k-2} +\ldots+ n(k-1)2 + n(k-1)$$

with $0 \le n(i) \le 1$ then the action of a generator g_j of G on V_n is given by multiplication by $(-1)^{n(j)}$.

The proof follows from the preceding results.

Before we can do anything non-trivial we shall need to order the subgroups of G in some natural way (we have already (essentially) ordered the subgroups of index 2 in G). Let $g_1, g_2, \ldots, g_k$ be a preferred base for G. Order the elements of G in the following way:

$$g_1 < g_2 <\ldots< g_k < g_1+g_2 < g_1+g_3 <\ldots< g_1+g_k < g_2+g_3 <\ldots$$
$$\ldots< g_{k-1}+g_k < g_1+g_2+g_3 <\ldots$$

first by "length" and then lexicographically.

Every subgroup H (of order 2^m) has a unique distinguished base $h_1 < h_2 <\ldots< h_m$ such that h_1 is the least non-zero element in H and for $i > 1$, h_i is the least element in H but not in $\langle h_1, h_2, \ldots, h_{i-1}\rangle$ the subgroup generated by $h_1, h_2, \ldots, h_{i-1}$. Using this distinguished base we may now order the subgroups of G, first by the order of the subgroup and then lexicographically on the distinguished base. So for example:

$$0 < \langle g_1\rangle < \langle g_2\rangle <\ldots< \langle g_k\rangle < \langle g_1,g_2\rangle <\ldots< \langle g_1,g_k\rangle <$$
$$\langle g_1,g_2+g_3\rangle <\ldots< \langle g_1,g_2+g_3+\ldots+g_k\rangle < \langle g_2,g_3\rangle <\ldots \ .$$

4.5.7 EXERCISE. Show that the two orderings given on the subgroups of G of index 2 actually coincide.

We now order the G slice types (essentially as described in section 1.7). We do this first by looking at the total dimension of the G slice type:

$$[H;V] < [K;W] \quad \text{if } \dim(V) < \dim(W) .$$

Next we look at the isotropy subgroup:

Suppose $\dim(V) = \dim(W)$, then $[H;V] < [K;W]$ if $H < K$

where we use the ordering on the subgroup already given. Finally we look at the lexicographical ordering of modules:

Suppose $\dim(V) = \dim(W)$, then $[H;V] < [H;W]$ if V precedes W with respect to the lexicographical ordering on H modules.

So for example, if $|H| = 4$ and if V_1, V_2, V_3 are the irreducible H modules we have

$$V_1 < V_2 < V_3$$
$$V_1 \oplus V_1 < V_1 \oplus V_2 < V_1 \oplus V_3 < V_2 \oplus V_2 < \ldots \quad .$$

We can thus denote the G slice types as

$$\rho_0, \rho_1, \rho_2, \ldots$$

with $\rho_i < \rho_j$ if and only if $i < j$, ($\rho_0 = [0,0]$).

Let F_j, as usual, be the set of G slice types

$$F_j = \{\rho_0, \rho_1, \ldots, \rho_j\} \qquad j \geq 0 .$$

Once again, it is not difficult to verify that each F_j is indeed a family of G slice types.

Let $S(H)$ denote the G slice types with H as isotropy subgroup.

Thus an element of $S(H)$ is of the form

$$\left[H; \sum_{\substack{L \subseteq H \\ 2}} \alpha(L) V(L) \right]$$

where $\alpha(L)$ is a non-negative integer and $V(L)$ is an irreducible H module with elements of H that belong to L acting trivially and all other elements acting by multiplication by -1. If K is a subgroup of H of index 2 then we have a function (*restriction*)

$$r = r_{H,K} : S(H) \to S(K)$$

given by

$$\left[H; \sum_{\substack{L \subseteq H \\ 2}} \alpha(L) V(L) \right] \longrightarrow \left[K; \sum_{\substack{L \subseteq H \\ 2 \\ L \neq K}} \alpha(L) V(K \cap L) \right] .$$

Note that $K \cap L$ is a subgroup of K of index 2. (Think of K, L as hyperplanes of codimension one in the $\mathbb{Z}/2$ vector space $H \cong (\mathbb{Z}/2)^r$ some r; then $K \cap L$ has codimension one in K and L if $K \neq L$.)

For each subgroup K of H of index 2 we also have the (*extension*) function

$$e = e_{K,H} : S(K) \to S(H)$$

defined by

$$\left[K; \sum_{\substack{L \subseteq K \\ 2}} \alpha(L) V(L) \right] \to \left[H, V(K) + \sum \alpha(L) V(\langle x \rangle \oplus L) \right]$$

where $H = \langle x \rangle \oplus K$ with x chosen minimally. Notice that $r_{H,K} e_{K,H} = 1$.

One reason for introducing the function e is given in the next result.

153

4.5.8 LEMMA. *If $\rho' = e(\rho)$ then*

$$f: N^G_n[\rho'] \to N^G_{n-1}[\rho]$$

defined by

$$E \to \nu_\rho S(E)$$

is an isomorphism.

PROOF. $f: N^G_*[\rho'] \to N^G_{*-1}[\rho]$ is an N_* module homomorphism. We shall define an N_* homomorphism

$$g: N^G_{*-1}[\rho] \to N^G_*[\rho']$$

which will be an inverse to f. Let ρ be written as

$$\rho = [K; \Sigma \alpha(L) V(L)]$$

then

$$\rho' = e(\rho) = [H; V(K) + \Sigma \alpha(L) V(\langle x \rangle \oplus L)]$$

where $H = \langle x \rangle \oplus K$ with x chosen minimally.

An element of $N^G_{n-1}[\rho]$ is a G vector bundle F over B with only K as isotropy subgroup in B. Thus the subgroup $\langle x \rangle$ acts freely on B and on F. Let L denote the line bundle over $B/\langle x \rangle$ associated to the double covering $B \to B/\langle x \rangle$. The induced action of G on $B/\langle x \rangle$ has isotropy subgroup H for every point in $B/\langle x \rangle$. The action of G on B induces in a natural way an action of G on L. In this way L is a G vector bundle of type $[H; V(K)]$. Since $\langle x \rangle$ acts freely on F, the quotient $F/\langle x \rangle$ is a G vector bundle over $B/\langle x \rangle$. It is of type $[H; \Sigma \alpha(L) V(\langle x \rangle \oplus L)]$ as can be easily seen by checking what happens in each fibre and recalling the definition of $V(L)$ etc. Define $g(F)$ to be the direct sum $L \oplus F/\langle x \rangle$ of the G vector bundles L and $F/\langle x \rangle$. The bundle $g(F)$ is of type ρ'. We leave it to the reader to verify that f and g are inverse to each other.

We now define subsets A_j, B_j, C_j of F_j in the following way:

$$A_0 = \{\rho_0\}, \quad B_0 = \emptyset, \quad C_0 = \emptyset.$$

Suppose that $A_{j-1}, B_{j-1}, C_{j-1}$ are defined. Consider ρ_j, there are two possibilities

(i) $\rho_j \neq e(\rho)$ for any $\rho \in A_{j-1}$
(ii) $\rho_j = e(\rho)$ for some $\rho \in A_{j-1}$.

In the first case let

$$A_j = A_{j-1} \cup \{\rho_j\}, \quad B_j = B_{j-1}, \quad C_j = C_{j-1}.$$

In the second case let ρ' be the minimal element of B_{j-1} such that $e(\rho') = \rho_j$, then define

$$A_j = A_{j-1} - \{\rho'\}$$
$$B_j = B_{j-1} \cup \{\rho_j\}$$
$$C_j = C_{j-1} \cup \{\rho'\}.$$

It is clear that

$$A_j \cup B_j \cup C_j = F_j.$$

In case (ii) above, the element ρ such that $e(\rho) = \rho_j$ is in fact unique.

4.5.9 LEMMA. *There is at most one G slice type $\rho \in A_{j-1}$ such that $e(\rho) = \rho_j$.*

We shall prove this and the next two results after some group theoretic results.

4.5.10 THEOREM. *Suppose that $\rho \in A_j$ and suppose that $[H;V]$ is a G slice type. If ρ is a G slice type of $G \times_H V$ then either $\rho = [H;V]$ or else $[H;V] \notin F_j$.*

This theorem implies that there is a well-defined homomorphism

$$\nu_i : N_*^G[F_j] \to N_*^G[\rho_i]$$

for all $\rho_i \in A_j$. This leads to the following result.

4.5.11 THEOREM.

$$N_*^G[F_j] \cong \bigoplus_{\rho_i \in A_j} N_*^G[\rho_i]$$

with the isomorphism being given by $\oplus \nu_i$.

The group theoretic results that we need give us criteria for when subgroups precede each other in the ordering of subgroups already given.

4.5.12 PROPOSITION. *Let $0 \subset K \subset H \subseteq G$ where $|K| = 2^{m-1}$ and $|H| = 2^m$. Let $\{g_1, g_2, \ldots, g_k\}$, $\{h_1, h_2, \ldots, h_m\}$, $\{k_1, k_2, \ldots, k_{m-1}\}$ be respectively, distinguished bases for G, H, K. Denote by r the least integer such that $k_i = h_i$ for all $i < r$. Then K is not contained in any predecessor of H if and only if $h_i = g_i$ for all $i \leq r$.*

PROOF. Assume that K is not contained in any predecessor of H. Suppose $h_i \neq g_i$ for some $i \leq r$, let j be the least such.

Then $g_j \notin H$, so $g_j \notin K$ and

$$K \subset \langle g_j \rangle \oplus K = H'.$$

Now, for all $i < j$ we have $k_i = h_i = g_i$ and so also we have $h'_i = g_i$ for all $i < j$ where $\{h'_1, h'_2, \ldots\}$ is the distinguished base of H'. We have $g_j \in H'$ and so $h'_j = g_j < h_j$. Thus $H' < H$ from the definition of ordering on the subgroups of G which contradicts the assumption and therefore $h_i = g_i$ for all $i \leq r$.

Conversely assume that $h_i = g_i$ for all $i \leq r$ and suppose that $K \subset H'$ where H' is of order 2^m and $H' \neq H$. Then $H' = \langle g \rangle \oplus K$ for some g – choose the least such g. Since $g \notin H$ it follows that $g > g_r$. Since $g_1, g_2, \ldots, g_{r-1} \in K$ we must have $h'_i = g_i$ for all $i < r$. By definition $k_r \neq h_r$. By hypothesis $h_r = g_r$ and so $g_r \notin K$. Thus, if $g_r \in H'$ then $H' = \langle g_r \rangle \oplus K \subseteq H$, which is a contradiction. So $g_r \notin H'$, so $h'_r > g_r$ and hence $H < H'$.

4.5.13 PROPOSITION. *Let $K, K' \subset H$ with K and K' not contained in any predecessor of H. Write $H = \langle x \rangle \oplus K$, $H = \langle x' \rangle \oplus K$ with x, x' chosen minimally. If $x \in K'$, $x' \in K$ and K precedes K' then $K \cap K'$ is not contained in any predecessor of K.*

PROOF. We have the following bases for H, K, K' and $L = K \cap K'$.

$H: g_1, g_2, \ldots, g_r, h_{r+1}, \ldots$

$K: g_1, g_2, \ldots, g_{r-1}, k_r, \ldots$ $\qquad k_r \neq g_r$

$K': g_1, g_2, \ldots, g_{s-1}, k'_s, \ldots$ $\qquad k'_s \neq g_s$

$L: g_1, g_2, \ldots, g_{s-1}, l_s, \ldots$

with $s < r$. This follows for H, K, K' from the previous proposition and for L by definition. We must have $x = g_r$ and $x' = g_s$, thus $g_r \in K'$, $g_s \in K$. Therefore

$$k'_s = g_t \quad \text{with} \quad s < t \leq r$$

(look at the distinguished base of K'). Also we must have $k_r > g_r$ since $k_r \neq g_r$. Let u be the least integer such that $l_i = k_i$ for all $i < t$. Then $u = s$ because if $l_s = k_s$ we get $l_s = g_s$ (because $k_s = g_s$ for $s < r$) but $g_s \notin K'$ and hence $g_s \notin L$, i.e., $l_s \notin L$. But for $i \leq u$ ($=s < r$) we have $k_i = g_i$ which by the previous proposition means that L is not contained in any predecessor of K.

To continue we shall need to define various subsets of $S(H)$, $H \subseteq G$. The functions

$$e = e_{K,H} : S(K) \longrightarrow S(H)$$

for $K \subseteq_2 H$ lead to a function

$$E : \bigcup_{\substack{K \subseteq_2 H \\ K \not\subseteq P(H)}} S(K) \to S(H)$$

where $K \not\subseteq P(H)$ means that K is not contained in any predecessor of H.

Let $\overline{S}(H)$ be defined by

$$\overline{S}(H) = S(H) - \text{Image} \ (E : \bigcup_{\substack{K \subseteq_2 H \\ K \not\subseteq P(H)}} S(K) \to S(H))$$

4.5.14 LEMMA. *Image*$(E) = $ *Image*$(\overline{E})$ *where* $\overline{E}$ *is the restriction* of E to $\bigcup\limits_{\substack{K \subsetneq H \\ K \not\in P(H)}} \overline{S}(K)$.

PROOF. $Im(\overline{E}) \subseteq Im(E)$ is clear.

Assume that $\rho \in Im(E)$, i.e., $\rho = e(\rho')$ where $\rho' \in S(K)$ with $K \subsetneq H$ and $K \not\in P(H)$. If $\rho' \not\in \overline{S}(K)$ then $\rho' = e(\rho'')$ for some $\rho'' \in S(L)$ with $L \subsetneq K$ and $L \not\in P(K)$ by definition of $\overline{S}(K)$. From Proposition 4.5.12 we have the following bases for H, K and L.

$H: g_1, g_2, \ldots, g_r, h_{r+1}, \ldots$
$K: g_1, g_2, \ldots, g_{r-1}, k_r, \ldots \qquad\qquad k_r \neq g_r$
$L: g_1, g_2, \ldots, g_{s-1}, l_s, \ldots \qquad\qquad l_s \neq g_s$

for some s, r with $s < r$. Furthermore

$$H = \langle g_r \rangle \oplus K, \qquad K = \langle g_s \rangle \oplus L$$

with g_r, g_s being the minimal possible choices in each case. Let $K' = \langle g_r \rangle \oplus L$, observe that $K' \not\in P(H)$ by Proposition 4.5.12 and that $\langle g_s \rangle \oplus K' = H$ and furthermore K precedes K'. Write ρ'' as

$$\rho'' = \left[L; \sum_{M \subsetneq L} V(M) \right]$$

thus

$$\rho' = e(\rho'') = \left[K; V(L) + \Sigma V(\langle g_s \rangle \oplus M) \right]$$

and

$$\rho = e(\rho') = \left[H; V(K) + V(\langle g_r \rangle \oplus L) + \Sigma V(\langle g_r \rangle \oplus \langle g_s \rangle \oplus M) \right].$$

However extending through K' gives

$$\rho''' = e_{L, K'}(\rho'') = \left[K'; V(L) + \Sigma V(\langle g_r \rangle \oplus L) \right]$$

in $S(K')$ and then extending to H

$$e_{K',H}(\rho''') = [H; V(K')+V(\langle g_s \rangle \oplus L)+\Sigma V(\langle g_s \rangle \oplus \langle g_r \rangle \oplus L)]$$

since $H = \langle g_s \rangle \oplus K'$.

Because $K = \langle g_s \rangle \oplus L$ and $K' = \langle g_r \rangle \oplus L$ we see that $\rho = e(\rho''')$ for some $\rho''' \in S(K')$. Since $L \subset K$ and K precedes K' then either $\rho''' \in \overline{S}(K')$ or $\rho''' = e(\rho^{iv})$ with $\rho^{iv} \in S(L')$ and $L' \subsetneq_2 K'$, $L' \not\subset P(K')$. In the latter case repeat the argument to obtain $\rho^v \in S(K'')$ where $K < K' < K''$ - continuing in this way we find some $\rho^{(2n+1)}$ in $\overline{S}(K^{(n)})$ such that $\overline{e}(\rho^{(2n+1)}) = \rho$ which completes the proof of the lemma.

4.5.15 LEMMA. *The function*

$$\overline{E}: \bigcup_{\substack{K \subsetneq_2 H \\ K \not\subset P(H)}} \overline{S}(K) \to S(H)$$

is injective.

PROOF. Suppose that

$$e(\rho) = e(\rho')$$

where $\rho \in S(K)$, $\rho' \in S(K')$ with $K < K'$, $K \subsetneq_2 H$, $K' \subsetneq_2 H$, $K \not\subset P(H)$, $K' \not\subset P(H)$. Under this assumption we shall show that $\rho \in e(\rho'')$ for some $\rho'' \in S(K \cap K')$. From Proposition 4.5.12 we see that $H = \langle g_r \rangle \oplus K$, $H = \langle g_s \rangle \oplus K'$ with g_r, g_s being minimal possible choices and $s < r$.

Write ρ and ρ' as

$$\rho = [K; \Sigma \alpha(L) V(L)]$$
$$\rho' = [K'; \Sigma \beta(M) V(M)].$$

Because $e(\rho) = e(\rho')$ we have

$$[H; V(K) \oplus \Sigma \alpha(L) V(\langle g_r \rangle \oplus K)] = [H; V(K') \oplus \Sigma \beta(M) V(\langle g_s \rangle \oplus M)]$$

which means that $g_s \in K$, $g_r \in K'$. By Proposition 4.5.13 this means that $K \cap K' \not\subset P(K)$ and $K = \langle g_s \rangle \oplus K \cap K'$ with g_s being the minimal possible choice.

Since $r_{H,K} e_{K,H} = 1$ we have

$$\rho = r_{H,K} e_{K,H}(\rho) = r_{H,K} e_{K',H}(\rho')$$
$$= [K; V(K \cap K') \oplus \Sigma \beta(H) V(\langle g_s \rangle \oplus M \cap K)]$$
$$= [K; V(K \cap K') \oplus \Sigma \beta(M) V(\langle g_s \rangle \oplus (M \cap K))]$$

since $g_s \in K$.

Now

$$r_{K, K \cap K'}(\rho) = [K \cap K'; \Sigma \beta(M) V((\langle g_s \rangle \oplus (M \cap K)) \cap (K \cap K'))]$$
$$= [K \cap K'; \Sigma \beta(M) V(M \cap K)]$$

since $M \subsetneq K'$ and $g_s \notin K'$.

Finally

$$e_{K \cap K', K}(r_{K, K \cap K'}(\rho)) = [K; V(K \cap K') \oplus \Sigma \beta(M) V(\langle g_s \rangle \oplus (M \cap K))]$$
$$= \rho.$$

But since $K \cap K' \subsetneq K$, $K \cap K' \not\subset P(K)$ and $\rho = e(\rho'')$ for some $\rho'' \in S(K \cap K')$ we get a contradiction (i.e., $\rho \notin \overline{S}(K)$) which proves the lemma.

Define $\overline{\overline{S}}(H)$ inductively by

$$\overline{\overline{S}}(H) = S(H) - \text{Image}(\overline{\overline{E}} : \bigcup_{\substack{K \subsetneq H \\ K \not\subset P(H)}} \overline{\overline{S}}(K) \to S(H))$$

with $\overline{\overline{S}}(0) = \{[0,0]\}$ and $\overline{\overline{E}}$ is the obvious restriction of E.

161

4.5.16 COROLLARY. *The function*

$$\overline{\overline{E}}: \bigcup_{\substack{K \subset H \\ K \not\subset P(H)}} \overline{\overline{S}}(K) \to S(H)$$

is injective.

PROOF. This follows from the previous two lemmas by induction.

4.5.17 PROOF OF LEMMA 4.5.9. We can now quite easily prove Lemma 4.5.9. Suppose that for $i < j$ there is at most one G slice type $\rho \in A_{i-1}$ such that $e(\rho) = \rho_i$. Consider ρ_j, and suppose that $e(\rho) = \rho_j$. Let $\rho_j \in S(H)$ and let $\rho \in S(K)$ with $K \subsetneq H$. If $K \subset P(H)$, say $K \subsetneq J$ with $J < H$ then

$$e_{K,J}(\rho) < \rho_j = e_{K,H}(\rho)$$

clearly. But then neither ρ nor $e_{K,J}(\rho)$ can belong to A_{j-1}. Thus $K \not\subset P(H)$ and from Corollary 4.5.16 we see that there is at most one ρ such that $e(\rho) = \rho_j$.

To prove Theorem 4.5.10 and 4.5.11 we still need another simple lemma.

4.5.18 LEMMA. *Let H be a subgroup of G and let K, L be subgroups of H of index 2. If $x \in H$ is the minimal element of H such that $\langle x \rangle \oplus K = H$ then $\langle x \rangle \oplus (L \cap K) \leq L$ with equality holding if and only if $x \in L$.*

PROOF. Let $L \cap K = J$. If $x \in L$ then $\langle x \rangle \oplus J = L$. Suppose therefore that $x \notin L$ and let $h_1, h_2, \ldots, h_n$ be a distinguished base for H. The element x must be h_r for some r because of the minimality condition of x. Thus we have $L = \langle y \rangle \oplus J$ for some y and $K = \langle y+h_r \rangle \oplus J$, (there are three subgroups that contain J and for which J has index 2, these are $\langle y \rangle \oplus J$, $\langle x \rangle \oplus J$, $\langle x+y \rangle \oplus J$). Note that $h_r < y$, otherwise

$$\langle y \rangle \oplus K = \langle y \rangle \oplus \langle y+h_r \rangle \oplus J = \langle h_r \rangle \oplus \langle y+h_r \rangle \oplus J = H$$

and this would contradict the minimality of $x = h_r$.

If the distinguished base of J is

$$j_1, j_2, \ldots, j_{n-2}$$

then $\langle h_r \rangle \oplus J$ has distinguished basis

$$j_1, j_2, \ldots, j_{s-1}, h_r, j'_{s+1}, \ldots, j'_{n-1}$$

while $\langle y \rangle \oplus J$ has distinguished basis

$$j_1, j_2, \ldots, j_{t-1}, y, j''_{t+1}, \ldots, j''_{n-1} \quad .$$

Since $y > h_r$ we have $t \geq s$ and hence

$$\langle h_r \rangle \oplus J < \langle y \rangle \oplus J = L$$

which proves the required result.

4.5.19 PROOF OF THEOREM 4.5.10. To prove Theorem 4.5.10, we suppose that ρ is a G slice type of $G \times_H V$ with $[H;V] \in F_j$. If $\rho \neq [H;V]$ then $\rho \in S(K)$ for some $K \subset H$ and there is a subgroup H' of H such that $K \subsetneq H' \subset H$. If $H' \neq H$ then

$$e_{K,H'}(\rho) < [H;V]$$

and so neither ρ nor $e_{K,H'}(\rho)$ belong to A_j. We may therefore assume that $H' = H$ and $K \subsetneq H$. Let us write $[H;V]$ as

$$[H;V] = \left[H; \sum_{L \underset{\neq}{\subseteq} H} \alpha(L)V(L)\right].$$

Since ρ is a G slice type of $G \times_H V$ and $\rho \neq [H;V]$ it follows that $\alpha(K) \neq 0$ and ρ is a G slice type of

$$G \times_K \left(\sum_{L \neq K} \alpha(L) V(L \cap K)\right)$$

(since $\rho \in S(K)$). But this is possible if and only if

$$\rho = \left[K; \sum_{L \neq K} \alpha(L) V(L \cap K)\right].$$

Now, $e_{K,H}(\rho)$ has the form

$$e_{K,H}(\rho) = \left[H; V(K) + \sum_{L \neq K} \alpha(L) V(\langle x \rangle \oplus (L \cap K))\right]$$

where x is the minimal element of G which gives $\langle x \rangle \oplus K = H$. From 4.5.18 we see that

$$e_{K,H}(\rho) \leq \left[H; V(K) + \sum_{L \neq K} \alpha(L) V(L)\right]$$

$$\leq [H;V]$$

and so neither ρ nor $e_{K,H}(\rho)$ belong to A_j which is a contradiction proving Theorem 4.5.10.

4.5.20 PROOF OF THEOREM 4.5.11. The proof of Theorem 4.5.11 is now not difficult. Consider the exact sequence

$$\ldots \to N_n^G[F_{j-1}] \to N_n^G[F_j] \xrightarrow{\nu_j} N_n^G[\rho_j] \xrightarrow{\partial_j} N_{n-1}^G[F_{j-1}] \to \ldots.$$

We prove 4.5.11 by induction, the result being trivially true for $j = 0$. Suppose that it is true for $j-1$, i.e., that

$$\oplus \nu_i : N_n^G[F_{j-1}] \to \bigoplus_{\rho_i \in A_{j-1}} N_n^G[\rho_i]$$

is an isomorphism. Consider the composite

$$N_n^G[\rho_j] \xrightarrow{\nu_i \partial_j} N_{n-1}^G[\rho_i] \quad .$$

If it is non-zero then ρ_i is a G slice type of $G \times_H V$, where $[H;V] = \rho_j$. Hence by 4.5.10 we have

$$\nu_i \partial_j \neq 0 \implies \rho_i \notin A_j \quad .$$

Thus, if $A_j = A_{j-1}$ then $\nu_i \partial_j = 0$ for all i such that $\rho_i \in A_{j-1}$ and hence the sequence becomes a short exact sequence

$$0 \to N_n^G[F_{j-1}] \to N_n^G[F_j] \xrightarrow{\nu_j} N_n^G[\rho_j] \to 0 \quad .$$

But short exact sequences of $\mathbb{Z}/2$ modules are split, using 4.5.10 we deduce that

$$\oplus \nu_i : N_n^G[F_j] \to \bigoplus_{\rho_i \in A_j} N_n^G[\rho_i]$$

is an isomorphism.

If on the other hand $A_j \neq A_{j-1}$ then there is precisely one i such that $\rho_i \in A_{j-1}$ but $\rho_i \notin A_j$ and $e(\rho_i) = \rho_j$. In this case, $\nu_i \partial_j$ is an isomorphism by 4.5.8 and we get a short exact sequence

$$0 \to N_{n+1}^G[\rho_j] \to N_n^G[F_{j-1}] \to N_n^G[F_j] \to 0 \quad .$$

The required result follows immediately.

4.5.21 REMARK. In the above we have not given an explicit splitting homomorphism. This is not necessary for the calculation of the generators of N_*^G except for the special case that $\rho_j = [G;V]$ and $\rho_j \neq e(\rho)$. An explicit splitting for these cases is given in the next section 4.6. It is probably not all that difficult to find explicit splitting homomorphisms for the other cases. As an example consider $G = (\mathbb{Z}/2)^2$. Let $G = \langle g_1, g_2 \rangle$, $G_1 = \langle g_1 \rangle$, $G_2 = \langle g_2 \rangle$, $G_3 = \langle g_3 \rangle$. The G slice types are of the form

$$[0,0]$$
$$[G_i; \tilde{R}^j] \quad i = 1,2,3 \quad j \geq 0$$
$$[G; V_1^a \times V_2^b \times V_3^c] \quad a,b,c \geq 0$$

where $V_i = V(G_i)$ $i = 1, 2, 3$.

We have the following isomorphisms

$$N_*^G[G_1; \tilde{R}^1] \cong N_{*-1}^G[0,0]$$
$$N_*^G[G; V_1 \times V_2^b] \cong N_{*-1}^G[G_1; \tilde{R}^b]$$
$$N_*^G[G; V_1^a \times V_2] \cong N_{*-1}^G[G_2; \tilde{R}^a]$$
$$N_*^G[G; V_1^a \times V_3] \cong N_{*-1}^G[G_3; \tilde{R}^a]$$

which we denote by f in each case. We use the following splitting homomorphisms

$$N_*^G[\rho_j] \to N_*^G[F_j]$$

(i) if $\rho_j = [G_1; \tilde{R}^b]$, $b \neq 1$ then

$$E \to RP(R \times E),$$

(ii) if $\rho_j = [G_2; \tilde{R}^a]$ or $[G_3, \tilde{R}^a]$ then

$$E \to S(f^{-1}(E)),$$

(iii) if $\rho_j = [G; V_1^a \times V_2^b \times V_3^c]$ and $\rho_j \neq e(\rho)$ then see section 4.6. This defines all the splitting homomorphisms that we need.

From Theorem 4.5.11 and 4.5.14-16 we have

4.5.22 COROLLARY.

$$N_n^G \cong \bigoplus_{\rho \in S^n(G)} N_n^G[\rho]$$

where $S^n(G)$ is given by

$$\{\rho \in S(G); \dim \rho \leq n, \rho \neq e(\rho') \text{ for any } \rho'\}$$

and the isomorphism is given by $\oplus \nu_\rho$.

4.5.23 COROLLARY.

$$N_n^G \cong \bigoplus_{n - \Sigma \alpha(H)} (\prod_H BO_{\alpha(H)})$$

where $\left[G; \sum_{H \subsetneq G} \alpha(H) V(H)\right] \in S^n(G)$.

This follows from 4.5.22 and the fact that if
$\rho = \left[G; \sum_{H \subsetneq G} \alpha(H) V(H)\right]$ then

$$B\Gamma(\rho) \cong \prod_H BO_{\alpha(H)} .$$

4.5.24 COROLLARY. *If M is a G manifold with no fixed points (under G) then M is a G boundary.*

This follows from 4.5.22 because if M has no fixed points then

$$\oplus \nu_\rho(M) = 0$$

for $\rho \in S^n(G)$ and so M is a G boundary.

To complete the proof of Theorem 4.5.1 we simply need an alternative description of when $\rho = e(\rho')$. This will be given shortly.

Recall that the non-trivial irreducible G modules are

$$\{V_i; 1 \leq i \leq 2^k-1\}$$

(Lemma 4.5.6). Let $G_1, G_2, \ldots$ be the corresponding subgroups of G of index 2

i.e., $V(G_i) = V_i$.

The following result is immediate.

4.5.25 LEMMA. *Let H be a subgroup of G of index 2. Then $H = G_l$ for some l, $2^a \leq l < 2^{a+1}$ if and only if $g_{k-a} \notin H$ and $g_i \in H$ for $i = 1, 2, \ldots, k-a-1$.*

In fact there are precisely 2^a, $(0 \leq a \leq k-1)$ subgroups of G of index 2 which satisfy

$$g_1, g_2, \ldots, g_{k-a-1} \in H, \quad g_{k-a} \notin H .$$

Furthermore, if $b > a$ then each of these subgroups precedes each of the 2^b subgroups K of G of index 2 which satisfy

$$g_1, g_2, \ldots, g_{k-b-1} \in K, \quad g_{k-b} \notin K.$$

4.5.26 COROLLARY. *Suppose $H = G_l$ with $2^a \leq l < 2^{a+1}$ for some $a, 0 \leq a < k$, then $\langle g_{k-a} \rangle \oplus H = G$ and g_{k-a} is the minimal possible choice.*

4.5.27 THEOREM. *Suppose that $(\alpha_0, \alpha_1, \alpha_2, \ldots)$ is a sequence of non-negative integers of length 2^k. Then*

$$\left[G; \sum_{n \geq 0} \alpha_n V_n\right] = e(\rho)$$

for some ρ if and only if there is an integer a such that

$\sum_{i=2^a}^{2^{a+1}-1} \alpha_i = 1$, say $\alpha_m = 1$ and $V_m \otimes V_n > V_n$ whenever $\alpha_n \neq 0$ $(n \neq m)$.

PROOF. First note that if
$$G_m = \langle x \rangle \oplus G_m \cap G_n$$
and
$$G_n = \langle y \rangle \oplus G_m \cap G_n$$
with x, y chosen minimally then
$$V_m \otimes V_n = V_l$$
where
$$G_l = \langle x+y \rangle \oplus G_m \cap G_n .$$

This follows quite easily from the description of $V_i = V(G_i)$ as the one dimensional real vector space with G_i acting trivially and all other elements acting non-trivially. Thus $G_m \cap G_n$ acts trivially on $V_m \otimes V_n$ and moreover so does the subgroup $\langle x+y \rangle$ since each of x and y act non-trivially on V_n, V_m respectively and trivially on V_m, V_n respectively. So
$$G_l = \langle x+y \rangle \oplus G_m \cap G_n$$
acts trivially on $V_m \otimes V_n$ and is itself a subgroup of G of index 2 thus $V_m \otimes V_n = V_l$.

Suppose that
$$[G; \Sigma \alpha_n V_n] = e(\rho)$$
$$= e([G_m; \Sigma \beta(L) V(L)])$$

$$= [G; V_m + \Sigma \beta(L) V(\langle g_{k-a} \rangle \oplus L)]$$

where $2^a \leq m < 2^{a+1}$. We see from Lemma 4.5.25 that

$$\sum_{i=2^a}^{2^{a+1}-1} \alpha_i = 1.$$ If $\alpha_n \neq 0$ $(n \neq m)$ then $\alpha_n = \beta(L)$ for some L and

$$G_n = \langle g_{k-a} \rangle \oplus L.$$

Also $V_m \otimes V_n = V_l$

$$G_l = \langle x + g_{k-a} \rangle \oplus L$$

assuming that $G_m = \langle x \rangle \oplus L$, (recall $L \subsetneq G_m$).

By Lemma 4.5.18 we have

$$G_n = \langle g_{k-a} \rangle \oplus L \leq \langle x + g_{k-a} \rangle \oplus L = G_l$$

(put $K = G_m$, $L = G_l$ in 4.5.18). If $g_{k-a} \in G_l$ then either $g_{k-a} = (x + g_{k-a}) + z$ or $g_{k-a} = z$ for some $z \in L$. In the first case this would show that $G_m = L$, while in the second that $G_n = L$ both contradictions. Thus $g_{k-a} \notin G_l$ and so we must have $G_l > G_m$. Thus $V_m \otimes V_n > V_n$.

Conversely suppose we are given $(\alpha_1, \alpha_2, \ldots)$ a $(2^k - 1)$-tuple of non-negative integers and an integer a such that

$$\sum_{i=2^a}^{2^{a+1}-1} \alpha_i = 1, \quad \text{say} \quad \alpha_m = 1$$

and furthermore that $V_m \otimes V_n > V_n$ whenever $\alpha_n \neq 0$ $(n \neq m)$. Consider

$$\rho = [G_m; \Sigma \alpha_n V(G_n \cap G_m)]$$

whence

170

$$e(\rho) = [G; V_m + \Sigma \alpha_n V(\langle g_{k-a} \rangle \oplus G_n \cap G_m)]$$

($2^a \leq m < 2^{a+1}$). The result will follow as soon as we show that $g_{k-a} \in G_n$ whenever $\alpha_n \neq 0$ because then

$$\langle g_{k-a} \rangle \oplus G_n \cap G_m = G_n$$

and

$$\sum_{i=2^a}^{2^{a+1}-1} \alpha_i = 1$$

by Lemma 4.5.25.

Let

$$G_m = \langle x \rangle \oplus L \quad \text{with} \quad 2^a \leq m < 2^{a+1}$$
$$G_n = \langle y \rangle \oplus L \quad \text{with} \quad 2^b \leq n < 2^{b+1}$$

and note that $b \neq a$ because of the condition $\sum_{i=2^a}^{2^{a+1}-1} \alpha_i = 1$.

If $\alpha_n \neq 0$ then by assumption

$$\langle x+y \rangle \oplus L > \langle y \rangle \oplus L = G_n.$$

If $b < a$ then $(k-b) > (k-a)$ and by Lemma 4.5.18 we must have $g_{k-a} \in G_n$. Suppose now that $b > a$ and suppose also that $g_{k-a} \notin G_n$. Since

$$\langle g_{k-b} \rangle \oplus G_n = G$$

we must have

$$g_{k-a} = g_{k-b} + g$$

for some $g \in G_n$, i.e., $g_{k-a} + g_{k-b} \in G_n$, or in other words

$$g_{k-a} + g_{k-b} = y + z$$

for some $z \in L$. Thus

$$g_{k-b} + y = g_{k-a} + z$$

and hence

$$\langle g_{k-b} + y \rangle \oplus L = \langle g_{k-a} \rangle \oplus L.$$

We know that $g_{k-b} \in G_m$ by Lemma 4.5.18 and the fact that $b > a$ and so $g_{k-b} \in L$ or $g_{k-b} = x$. The first being impossible so $g_{k-b} = x$. Hence by assumption we conclude that

$$\langle g_{k-a} \rangle \oplus L = \langle g_{k-b} + y \rangle \oplus L = \langle x+y \rangle \oplus L > \langle y \rangle \oplus L = G_n$$

which contradicts Lemma 4.5.25. This completes the proof of Theorem 4.5.27.

To complete the proof of Theorem 4.5.1 we observe that there are one-to-one correspondences

$$\{x_m \in G\} \leftrightarrow \{m; 0 \leq m \leq 2^k - 1\} \leftrightarrow \{G_m \subsetneq G\}.$$

Each integer determines an element of G via its 2-adic expansion and each element of G determines a unique subgroup of index 2 in G (the subspace of $(\mathbb{Z}/2)^k$ orthogonal to the given vector = element of G). We have already defined $m*n$, the sum $x_m + x_n$ is given. The subgroup G_{m*n} is obtained from G_m and G_n in the following way

$$G_m = \langle x \rangle \oplus G_m \cap G_n$$
$$G_n = \langle y \rangle \oplus G_m \cap G_n$$
$$G_{m*n} = \langle x+y \rangle \oplus G_m \cap G_n$$

where x and y are chosen minimally. The result in Theorem 4.5.25 may now be stated as:

Suppose $(\alpha_1, \alpha_2, \ldots)$ is a sequence of non-negative integers. Then $[G; \Sigma \alpha_n V_n] = e(\rho)$ for some ρ if and only if there exists an integer a such that $\sum_{i=2^a}^{2^{a+1}-1} \alpha_i = 1$, say $\alpha_m = 1$, and $m*n > n$ whenever $\alpha_n \neq 0$ ($n \neq m$).

Theorem 4.5.1 follows immediately.

4.6 GENERATORS OF $(\mathbb{Z}/2)^k$ BORDISM

We shall show that the generators of N_*^G, for $G = (\mathbb{Z}/2)^k$ are of the form

$$\frac{\prod_{i=0}^{k-1} S(R \times \xi_{J(i)})}{T^k}$$

where $J(i) = (j(i,0), j(i,1), \ldots, j(i, 2^i - 1))$ is a 2^i-tuple of non-negative integers, $\xi_{J(i)}$ has its usual meaning as a product of canonical line bundles over real projective spaces and $S(-)$ denotes the sphere bundle. The action of the group $T^k = (\mathbb{Z}/2)^k$ on $\Pi S(R \times \xi_{J(i)})$ depends upon $(J_0, J_1, \ldots, J_{k-1})$. The precise action of G on such a manifold will be given shortly.

Recall that the non-trivial irreducible G modules are

$$\{V_i; 1 \leq i \leq 2^k - 1\}$$

where if $g_1, g_2, \ldots, g_k$ is a base for G and if m is written in its 2-adic form as

$$m = m_1 2^{k-1} + m_2 2^{k-2} + \ldots + m_k$$

then g_i acts on V_m by multiplication by $(-1)^{m_i}$.

173

4.6.1 THEOREM. *An N_* base for N_*^G, $G = (\mathbb{Z}/2)^k$ is given by the following G manifolds.*

$$\left(\prod_{i=0}^{k-1} S\left(W(2^{i-1}) \times \prod_{j=2^i}^{2^{i+1}-1} \prod_{l=1}^{\alpha_j} \xi(n_{j,l}) \otimes V_j \otimes W_{q(\alpha,j)} \right) \right) / T^k$$

where $\alpha = (\alpha_1, \alpha_2, \ldots) \in S = \bigcup_{n \geq 0} S^n$ and

$$n_{j,1} \geq n_{j,2} \geq \ldots \geq n_{j,\alpha_j} \geq 0$$

is a sequence of non-negative integers.

$W_{q(\alpha,j)}$ and $W(2^{i-1})$ are one dimensional real vector spaces with an action of $T^k = (\mathbb{Z}/2)^k$. If T^k has a base $t_1, t_2, \ldots, t_k$ then T^k acts on $W_{q(\alpha,j)}$ and $W(2^{i-1})$ as follows:

t_i acts on $W_{a(\alpha,j)}$ by multiplication by -1 if either

(a) $2^{i-1} \leq j < 2^i$, or

(b) $j \geq 2^i$ and $\sum_{l=2^{i-1}}^{2^i-1} \alpha_l = 1$ with $\alpha_m = 1$ (say) and $m*j < j$,

otherwise t_i acts by multiplication by $+1$.

t_i acts on $W(2^{j-1})$ by multiplication by -1 if $i = j$ and otherwise by multiplication by $+1$.

The manifolds above are fibre bundles over

$$\prod_{i=0}^{k-1} \prod_{j=2^i}^{2^{i+1}-1} \prod_{l=1}^{\alpha(j)} RP^{n(j,l)}$$

with fibre

$$\left(\prod_{i=0}^{k-1} S\left(W(2^{i-1}) \times \prod_{j=2^i}^{2^{i+1}-1} (V_j^{\alpha(j)} \otimes W_{q(\alpha,j)}) \right) \right) / T^k \quad .$$

4.6.2 REMARK. If there does *not* exist an i such that

$$\sum_{l=2^i}^{2^{i+1}-1} \alpha_l = 1$$

then the G manifold defined above is precisely

$$\prod_{i=0}^{k-1} RP\left(R \times \prod_{j=2^i}^{2^{i+1}-1} \prod_{l=1}^{\alpha(j)} \xi(n_{j,l}) \otimes V_j \right).$$

The reader may like to check that for the case $G = (\mathbb{Z}/2)^2$ the manifolds in the theorem are

(i) $RP(R \times (\xi_J \otimes V_1)) \times RP(R \times (\xi_K \otimes V_2) \times (\xi_L \otimes V_3))$

where $J = (j_1, j_2, \ldots, j_a)$ with $j_1 \geq j_2 \geq \ldots \geq j_a \geq 0$,

$K = (k_1, k_2, \ldots, k_b)$ with $k_1 \geq k_2 \geq \ldots \geq k_b \geq 0$,

$L = (l_1, l_2, \ldots, l_c)$ with $l_1 \geq l_2 \geq \ldots \geq l_c \geq 0$,

and $a \neq 1$, $b+c \neq 1$.

(ii) $(S(R \times (\xi_j \otimes V_1)) \times RP(R \times (\xi_K \otimes V_2) \times (\xi_L \otimes V_3)))/(\mathbb{Z}/2)$

where $j \geq 0$, $K = (k_1, k_2, \ldots, k_b)$ with $k_1 \geq k_2 \geq \ldots \geq k_b \geq 0$, $L = (l_1, l_2, \ldots, l_c)$ with $l_1 \geq l_2 \geq \ldots \geq l_c \geq 0$ and $c \neq 0$, $b+c \neq 1$.

The action of $\mathbb{Z}/2$ is given by

$$(w, [x,y,z]) \to (-w, [x,y,-z])$$

where $w \in S(R \times (\xi_j \otimes V_1))$ and $[x,y,z] \in RP(R \times \ldots)$ with x coming from R, y coming from $(\xi_K \otimes V_2)$ and z coming from $(\xi_L \otimes V_3)$. (Recall that $S = \{(a,b,c); a \neq 1, b+c \neq 1\} \cup \{(1,b,c); c \neq 0, b+c \neq 1\}$.)

Let us look at the slice types occuring in a typical fibre

$$\left(\prod_{i=0}^{k-1} S(W(2^{i-1}) \times \prod_{j=2^i}^{2^{i+1}-1} V_j^{\alpha(j)} \otimes W_{q(\alpha,J)}) \right) / T^k .$$

In fact we need only look at those slice types with G as isotropy subgroup. Suppose then that we are given $\alpha = (\alpha(1), \alpha(2), \ldots, \alpha(2^k-1)) \in S$ and suppose also that the condition

$$\sum_{j=2^i}^{2^{i+1}-1} \alpha_i = 1$$

is satisfied precisely for $i = i_1, i_2, \ldots, i_l$ and that $\alpha_m = 1$ for $m = m_1, m_2, \ldots, m_l$

$$2^{i(j)} \leq m_j < 2^{i(j)+1}$$

Thus t_i (a base element of T^k) acts on $W_{q(\alpha,j)}$ by multiplication by

$\quad -1 \quad$ if $2^{i-1} \leq j < 2^i$

$\quad -1 \quad$ if $i = i_n$ some $n = 1, 2, \ldots, l$ with $j \geq 2^i$ and $m_n * j < j$

$\quad +1 \quad$ otherwise .

We shall denote a point in the fibre

$$F = (\Pi S(W(2^{i-1})) \times \Pi V_j^{\alpha(j)} \otimes W_{q(\alpha,j)}) / T^k$$

by

$$\prod_{i=0}^{k-1} (x(i,0), x(i,1), \ldots, x(i,2^i))$$

where $x(i,0) \in W(2^{i-1})$, $x(i,j) \in V_l^{\alpha(l)}$ ($l = 2^i + j - 1$) and

$$(x(i,0), x(i,1), \ldots, x(i,2^i)) \in S^{\Sigma \alpha(j)} \quad .$$

Such a point is fixed by the action of G if and only if it is of the form

$$\prod_{i=0}^{k-1} (0, 0, \ldots, 0, x(i, n(i)), 0, \ldots, 0)$$

with $n(i)$ depending on i.

The point

$$\prod_{i=0}^{k-1} (x(i,0), 0, \ldots, 0)$$

where $x(i,0) = \pm 1$ (this represents one point in F) has slice type

$$\left[G; \prod_i \prod_j V_j^{\alpha(j)} \right] = \rho \quad \text{say}.$$

4.6.3 THEOREM. *The maximal slice type of points in F is precisely ρ. Furthermore there is only one point with this slice type.*

By maximal we mean maximal with respect to the ordering of G slice types given in the previous section (4.5).

PROOF. The slice type of the point

$$\prod_i (0, 0, \ldots, 0, x(i, n(i)), 0, \ldots, 0)$$

is

$$\left[G; \prod_{i} \prod_{\substack{j=0 \\ j \neq n(i)}}^{2^i} (V_{i,j}^{\alpha(i,j)} \otimes (\bigotimes_{a \in A(i,j)} V_{m(a)}) \otimes V_{i,n(i)} \otimes\right.$$

$$\left.(\bigotimes_{b \in A(i,n(i))} V_{m(b)}))\right]$$

$$= \rho(n_0, n_1, \ldots, n_{k-1}) \quad \text{say}$$

where

$$V_{i,j} = \begin{cases} V_{2^i+j-1} & \text{if } j \neq 0 \\ V_0 = R & \text{if } j = 0 \end{cases}$$

and

$$A(i,j) = \{a; 1 \leq a \leq l, \ 2^b + n(b) - 1 = m(a) \text{ for some}$$
$$b < i \text{ with } n(b) \neq 0 \text{ and } V_{m(a)} \otimes V_{i,j} < V_{i,j}\}.$$

The above calculation is quite easy, it follows from the definition of the action of T^k on $\Pi S(--)$.

4.6.4 REMARK. From the definition of S^n it follows quite easily that $A(i, n(i)) \neq \emptyset$ for at least one i.

Now $\dim \rho(n_0, n_1, \ldots, n_{k-1}) < \dim \rho$ unless $\alpha(i, n(i)) = 1$ for $i = 0, 1, \ldots, k-1$. Suppose that $\alpha(i, n(i)) = 1$ for $i = 0, 1, \ldots, k-1$. We have the following

4.6.5. (i) If $jk \neq 0$ then $V_{i,j} \otimes V_{i,k} = V_{l,m}$ for some $l < i$ and some m.

(ii) If $i < k$ and $jl \neq 0$ then $V_{i,j} \otimes V_{k,l} = V_{k,m}$ for some $m \leq l$.

(iii) If $i < k < m$, $jln \neq 0$, $V_{i,j} \otimes V_{m,n} < V_{m,n}$ and if $V_{k,l} \otimes V_{m,n} < V_{m,n}$ then

$$V_{i,j} \otimes V_{k,l} \otimes V_{m,n} < V_{m,n}.$$

Each of the statements above has either been proved previously or can be quite easily deduced from the fact that $V_a \otimes V_b = V_{a*b}$ and then looking at $a*b$.

Let V be given by

$$V = V_{i,j} \otimes (\bigotimes_{a \in A(i,j)} V_{m(a)}) \otimes V_{i,n(i)} \otimes (\bigotimes_{b \in A(i,n(i))} V_{m(b)}).$$

From 4.6.5(i), (ii) and (iii) we deduce that

4.6.6   (a)   If $j \neq 0$ and $n(i) \neq 0$ then $V < V_{i,j}$.

        (b)   If $j = 0$ then $V \leq V_{i,n(i)}$ with equality only if $A(i,n(i)) = \emptyset$.

        (c)   If $n(i) = 0$ then $V \leq V_{i,j}$ with equality only if $A(i,j) = \emptyset$.

It follows that $\rho(n_0, n_1, \ldots, n_{k-1}) \leq \rho$, with equality only if for each i either $n(i) = 0$ or $A(i,n(i)) = \emptyset$. However, $A(i,n(i)) \neq \emptyset$ for at least one i, therefore equality holds only if $n(i) = 0$ for $i = 0, 1, \ldots, k-1$. But then $\rho(0,0,\ldots,0)$ and ρ are the slice types of the same point. This proves Theorem 4.6.3.

Given $\rho = [G; \prod_{i=0}^{k-1} \prod_{j=2^i}^{2^{i+1}-1} V_j^{\alpha(j)}]$ we can therefore define an N_* homomorphism

$$q_\rho : N_*^G[\rho] \to N_*^G$$

which on the basis elements of $N_*^G[\rho]$ is defined by

$$\prod_{i=0}^{k-1} \prod_{j=2^i}^{2^{i+1}-1} \prod_{l=1}^{\alpha(j)} \xi(\eta_{j,l}) \otimes V_j \;\to\; (\prod_{i=0}^{k-1} S(\ldots))/T^k$$

where $\Pi S(\ldots)/T^k$ is the manifold defined in Theorem 4.6.1. From Theorem 4.6.3 we see that

$$\nu_\rho q_\rho = 1$$

$$\nu_{\rho'} q_\rho = 0 \quad \text{if } \rho' > \rho .$$

Thus Theorem 4.6.1 follows now from Theorem 4.5.22.

4.7 GENERATORS FOR G BORDISM, G GENERAL

Although we cannot at present find generators of N_*^G for all groups G, we shall at least state a conjecture concerning the algebraic side of N_*^G.

Suppose that the G slice types are totally ordered somehow. Usually this ordering is as described in 1.7. In other words first we look at the dimension of the slice type:

$$[H;V] < [K;W] \quad \text{if } \dim(V) < \dim(W).$$

Secondly, we look at the isotropy subgroup:

suppose $\dim(V) = \dim(W)$, then $[H;V] < [K;W]$ if $H < K$ where the subgroups of G are ordered in some way satisfying the condition that if H is a subgroup of K then $H < K$. Finally we order the irreducible H modules for all subgroups H of G (as remarked in 1.7, an ordering of the subgroups of G gives (almost) an ordering of the irreducible H modules). This induces the lexicographical ordering on the H modules and we have:

Suppose $\dim(V) = \dim(W)$, then $[H;V] < [H;W]$ if $V < W$.

Call the G slice types

$$\rho_0, \rho_1, \rho_2, \ldots$$

with the obvious condition that if $i < j$ then $\rho_i < \rho_j$. Observe that if we are given $\rho_j = [H;V]$ and if ρ_i is a slice type of $G \times_H V$ then $i \le j$. Thus the set F_j given by

$$F_j = \{\rho_0, \rho_1, \ldots, \rho_j\}$$

is a family of G slice types. Define, inductively, subsets A_j, B_j of F_j as follows. Let A_0, B_0 be given by

$$A_0 = \{\rho_0\} \qquad B_0 = \emptyset \quad .$$

Suppose that A_{j-1}, B_{j-1} are defined ($j \ge 1$) and consider $\rho_j = [H;U]$. There are two possibilities.

(i) There does not exist a subgroup K of index 2 in H such that $U = V_K \times W$ and $V_K \not< W$.

By V_K we mean the H module whose underlying space is R and with H action given by

$$hx = \begin{cases} -x & \text{if } h \notin K \\ x & \text{if } h \in K \end{cases}$$

(ii) There exists a subgroup K of index 2 in H such that $U = V_K \times W$ and $V_K \not< W$.

In case (i) we define $A_j = A_{j-1}$, $B_j = B_{j-1}$. In case (ii) consider the following set X.

$$X = \{ [K;W|K] \in A_{j-1}; [H;U] = [H; V_K \times W], K \underset{2}{\subsetneq} H, V_K \not< W \} \quad .$$

(Notice that for the cases dealt with in the previous sections

$X = \emptyset$ or X consists of a single element.) If $X = \emptyset$ then let $A_j = A_{j-1}$, $B_j = B_{j-1}$. If $X \neq \emptyset$ then let $r[H;U]$ be the minimal element of X and define

$$A_j = A_{j-1} - \{r[H;U]\}$$
$$B_j = B_{j-1} \cup \{[H;U]\} \ .$$

Notice that the following always holds:

$$A_j \cup B_j \cup \{r(\rho); \rho \in B_j\} = F_j.$$

4.7.1 CONJECTURE. $N_*^G[F_j]$ *is isomorphic as an* N_* *module to the direct sum of*

$$\bigoplus_{\rho \in A_j} N_*^G[\rho]$$

$$\bigoplus_{\rho \in B_j} ker(\nu\partial : N_*^G[\rho] \to N_{*-1}^G[r(\rho)])$$

$$\bigoplus_{\rho \in B_j} N_*^G[r(\rho)]/(im(\nu\partial : N_{*+1}^G[\rho] \to N_*^G[r(\rho)])).$$

This conjecture is indeed true for many groups G. In the previous sections we have shown this to be true for the case that G is of odd order, $G = \mathbb{Z}/2^r$ and $G = (\mathbb{Z}/2)^k$. It is not difficult to show that the conjecture is also true for groups which have a cyclic sylow 2-subgroup, i.e., groups of the form $G \times \mathbb{Z}/2^r$ with G of odd order. We illustrate this for the case $r = 1$, as well as finding the generators of $N_*^{G \times \mathbb{Z}/2}$, G odd order.

The calculation of the generators of $N_*^{G \times \mathbb{Z}/2}$ is not very difficult so we shall only mention the essential points leaving the finer details to the reader. From now let G be a finite abelian group of odd order. The subgroups of $G \times \mathbb{Z}/2$ are of the form $H \times \mathbb{Z}/2$, H where H is a subgroup of G. If the order of H is $(2k+1)$ then the irreducible H modules are

$$R, V_1, V_2, \ldots, V_k$$

while the irreducible $H \times \mathbb{Z}/2$ modules are

$$R, V_1, V_2, \ldots, V_k$$
$$\tilde{R}, \tilde{V}_1, \tilde{V}_2, \ldots, \tilde{V}_k$$

where $\tilde{R}$ has $H \times \mathbb{Z}/2$ action given by

$$(h, \pm 1)x = \pm x$$

and $\tilde{V}_i = V_i \otimes \tilde{R}$. We see that a typical $G \times \mathbb{Z}/2$ slice type is of the form

$$\left[H; \sum_{i=1}^{k} \alpha_i V_i \right]$$

or

$$\left[H \times \mathbb{Z}/2; \alpha \tilde{R} + \sum_{i=1}^{k} \alpha_i V_i + \sum_{i=1}^{k} \tilde{\alpha}_i \tilde{V}_i \right] \quad .$$

Order the subgroups of $G \times \mathbb{Z}/2$ so that if $A \subseteq B$ then $A \leq B$. Order the non-trivial irreducible $H \times \mathbb{Z}/2$ modules (and H modules) by

$$\tilde{R} < V_1 < V_2 < \ldots < V_k < \tilde{V}_1 < \tilde{V}_2 < \ldots < \tilde{V}_k .$$

By the method described at the beginning of this section we can now totally order our $G \times \mathbb{Z}/2$ slice types and call them

$$\rho_0, \rho_1, \rho_2, \ldots$$

and define families of $G \times \mathbb{Z}/2$ slice types F_j by

$$F_j = \{\rho_0, \rho_1, \ldots, \rho_j\} .$$

We can then proceed to define subsets A_j, B_j. It is clear that $\rho \in B_j$ only if it is of the form

$$\left[H \times \mathbb{Z}/2; \tilde{R} + \sum_{i=1}^{k} \alpha_i V_i \right]$$

and $r(\rho)$ is then

$$\left[H; \sum_{i=1}^{k} \alpha_i V_i \right] .$$

It is trivial to show that $\nu \partial : N^G_*[\rho] \to N^G_*[r(\rho)]$ is an isomorphism for $\rho \in B_j$. Also it is quite easy to show that

$$N^G_*[F_j] = \bigoplus_{\rho \in A_j} N^G_*[\rho] .$$

In fact, elements of A_j are of the form

$$\rho = \left[H \times \mathbb{Z}/2; \alpha \tilde{R} + \sum_{i=1}^{k} \alpha_i V_i + \sum_{i=1}^{k} \tilde{\alpha}_i \tilde{V}_i \right]$$

with $\alpha \neq 1$ or $\sum_{i=1}^{k} \tilde{\alpha}_i \neq 0$ or of the form

$$\rho = \left[H; \sum_{i=1}^{k} \alpha_i V_i \right]$$

with $\left[H \times \mathbb{Z}/2; \tilde{R} + \sum_{i=1}^{k} \alpha_i V_i \right] \notin F_j$.

So, if $\rho_j \in A_j$ then we can define a splitting homomorphism

$$q : N^G_*[\rho_j] \to N^G_*[F_j]$$

by the following: let $E \in N_*^G[\rho_j]$

(i) If $\rho_j = [H; \sum \alpha_i V_i]$

$$q(E) = RP(R \times (E_1, E_2, \ldots, E_k)).$$

(ii) If $\rho_j = [H \times \mathbb{Z}/2; \alpha\tilde{R} + \sum \alpha_i V_i + \sum \tilde{\alpha}_i \tilde{V}_i]$ and $\alpha \neq 1$ then

$$q(E) = RP(R \times (E_0, E_1, E_2, \ldots, E_k, \tilde{E}_1, \tilde{E}_2, \ldots, \tilde{E}_k)).$$

(iii) If $\rho_j = [H \times \mathbb{Z}/2; \tilde{R} + \sum \alpha_i V_i + \sum_{i=j}^{k} \tilde{\alpha}_i \tilde{V}_i]$ with $\tilde{\alpha}_j \neq 0$ then

$$q(E) = RP(R \times (E_0 \oplus \tilde{E}_j, E_1, \ldots, E_k, \tilde{E}_{j+1}, \ldots, \tilde{E}_k)).$$

The subbundles $E_0, E_1, \ldots, E_k, \tilde{E}_1, \ldots, \tilde{E}_k$ of E are defined in the obvious way. By defining slightly different splittings, directly on the generators of $N_*^G[\rho_j]$ we obtain the next result.

4.7.2 THEOREM. $N_*^{G \times \mathbb{Z}/2}$ *is multiplicatively generated as an* N_* *module by* G/H, $(H \subseteq G)$ *and the following:*

$RP(R \times (\eta_j \otimes V))$

$RP(R \times (\eta_j \otimes V \otimes \tilde{R}))$

$RP(R \times (\xi_I \otimes \tilde{R}))$

$RP(R \times (\xi_i \otimes \tilde{R}) \times (\eta_j \otimes V \otimes \tilde{R}))$

where V *is a non-trivial irreducible* G *module and* $\tilde{R}$ *denotes the non-trivial irreducible* $\mathbb{Z}/2$ *module.*

The integers i, j are non-negative and I is a non-increasing sequence of non-negative integers of length not equal to one.

The calculation of the generators of $N_*^{G\times \mathbb{Z}/2^r}$, G of odd order, is not difficult and the reader is urged to do the next exercise.

4.7.3 EXERCISE. If G is a group of odd order show that $N_*^{G\times \mathbb{Z}/2^r}$ is *multiplicatively* generated as an N_* module by G/H, $(H \subseteq G)$ and the following:

$$X \times_{\mathbb{Z}/2^s} RP(R\times \xi_I \otimes \tilde{R})$$

$$RP(R\times \eta_j \otimes V)$$

$$X \times_{\mathbb{Z}/2^s} RP(R\times (\xi_i \otimes \tilde{R})\times (\eta_j \otimes V))$$

where $X = \mathbb{Z}/2^r$ or S^1 (unless $s=r$ in which case $X = \mathbb{Z}/2^r$) and V runs through the non-trivial irreducible $G \times \mathbb{Z}/2^r$ modules of dimension 2.

As a hint note that the irreducible $H \times \mathbb{Z}/2^s$ modules are of the form

$$R, V_1, V_2, \ldots, V_k$$
$$\tilde{R}, V_1\otimes\tilde{R}, V_2\otimes\tilde{R}, \ldots, V_k\otimes\tilde{R}$$
$$V_i \otimes W_j \quad i = 0, 1, \ldots, 2k, \quad j = 1, 2, \ldots, 2^{s-1}$$

where $V_0, V_1, \ldots, V_{2k}$ are the irreducible complex H modules, $R, V_1, V_2, \ldots, V_k$ are the irreducible real H modules ($|H| = 2k+1$) and $R, \tilde{R}, W_1, W_2, \ldots$ are the irreducible $\mathbb{Z}/2^s$ modules. In fact if W_j denotes the complex numbers with the generator of $\mathbb{Z}/2^s$ acting by multiplication by $\exp(2\pi i j/2^s)$ then we can rewrite the

$H \times \mathbb{Z}/2^s$ modules in the following more convenient form:

$$R, V_1, V_2, \ldots, V_k$$

$$\tilde{R}, V_1 \otimes \tilde{R}, V_2 \otimes \tilde{R}, \ldots, V_k \otimes \tilde{R}$$

$$V_i \otimes W_j \quad i = 0, 1, \ldots, 2k, \quad j = 1, 2, \ldots, 2^{s-2}$$

$$V_i \otimes W_j \otimes \tilde{R} \quad i = 0, 1, \ldots, 2k, \quad j = 1, 2, \ldots, 2^{s-2}-1.$$

Note that Exercise 4.7.3 gives generators of N_*^G when G is a finite cyclic group.

4.8 GENERATORS OF S^1 BORDISM

Although S^1 is not a finite group, it is abelian and the methods developed enable us to quite easily write down the generators of S^1 bordism. Note that the non-trivial irreducible S^1 modules are

$$V_1, V_2, V_3, \ldots$$

where V_j denotes the complex numbers with $t \in S^1 \subseteq C$ acting by multiplication by t^j.

4.8.1 THEOREM. N_*^G, $(G = S^1)$ is a free N_* module generated by

(i) $\quad S^1 \times_{\mathbb{Z}/2} RP(R \times (\xi_I \otimes \tilde{R}))$

(ii) $\quad S^1 \times_{\mathbb{Z}/2m} RP(R \times (\xi_I \otimes \tilde{R})) \times \prod_{i=1}^{l} RP(R \times (\eta_{j(i)} \otimes V_{k(i)}))$

(iii) $\quad S^1 \times_{\mathbb{Z}/2m} RP(R \times (\xi_a \otimes \tilde{R}) \times (\eta_b \otimes V_j)) \times \prod_{i=1}^{l} RP(R \times (\eta_{j(i)} \otimes V_{k(i)}))$

(iv) $\quad \prod_{i=1}^{l} RP(R \times (\eta_{j(i)} \otimes V_{k(i)}))$

where I is a sequence of non-increasing non-negative integers of length greater or equal to 2. The integers $l, j(i), k(i), a, b$ and j are non-negative while $m \geq 2$. In (ii), (iii) and (iv) if $V_{k(q)} = V_{k(r)}$ for $q > r$ then $j(q) \geq j(r)$. In (ii) $1 \leq k(i) < m$ while in (iii) $m/2 \leq j < m$ and $1 \leq k(i) \leq j$. Furthermore, in (iii), if $j = m/2$ then $a > 0$ and b is even.

We shall merely give an outline of the proof of the theorem. The reader should have very few problems in filling in the details.

4.8.2 The S^1 slice types are

$$\left[\mathbb{Z}/(2m+1); \prod_{i=1}^{l} V_{k(i)} \right] \qquad l \geq 0, \ 1 \leq k(i) \leq m, \ m \geq 0$$

$$\left[\mathbb{Z}/2m; \mathbb{R}^j \times \prod_{i=1}^{l} V_{k(j)} \right] \qquad j \geq 0, \ l \geq 0, \ 1 \leq k(i) \leq m-1, \ m \geq 1$$

$$\left[S^1; \prod_{i=1}^{l} V_{k(i)} \right] \qquad l \geq 0, \ k(i) \geq 1.$$

The action of $\mathbb{Z}/m$ on V_k is the action induced from the natural inclusion $\mathbb{Z}/m \subset S^1$. If $m = 0$ then by $\left[\mathbb{Z}/(2m+1); \prod_{i=1}^{l} V_{k(i)} \right]$ we mean $[1; 0]$.

4.8.3 From section 1.5 we have that

$$N_*^G [\mathbb{Z}/m; V] \cong N_{*-1-dim\ V}(B\Gamma(\mathbb{Z}/m; V))$$

where the extra 1 appears because $\dim S^1 = 1$. Also,

$$N_*^G [S^1; V] \cong N_{*-dim\ V}(B\Gamma(S^1; V)) \ .$$

4.8.4 From section 3.2 we easily see that

$$B\Gamma(\mathbb{Z}/(2m+1); \prod_{i=1}^{m+1} V_i^{d(i)}) \cong B(S^1/(\mathbb{Z}/(2m+1))) \times \Pi BU_{d(i)}$$

$$\cong BS^1 \times \Pi BU_{d(i)}$$

$$B\Gamma(\mathbb{Z}/2m; \tilde{R}^j \times \prod_{i=1}^{m} V_i^{d(i)}) \cong B((S^1 \times O_j)/(\mathbb{Z}/2m)) \times \Pi BU_{d(i)}$$

$$\cong B((S^1/(\mathbb{Z}/m)) \times_{\mathbb{Z}/2} O_j) \times \Pi BU_{d(i)}$$

$$\cong B(S^1 \times_{\mathbb{Z}/2} O_j) \times \Pi BU_{d(i)} \quad .$$

Furthermore, $B(S^1 \times_{\mathbb{Z}/2} O_{2j+1}) \cong B(S^1 \times SO_{2j+1})$.

$$B\Gamma(S^1; \prod_{i \geq 1} V_i^{d(i)}) \cong \Pi BU_{d(i)} \quad .$$

The next result we need is the generators of $N_*^G[H;U]$ for the various S^1 slice types $[H;U]$. From chapter 3 we know the generators of $N_*(BU_d)$. Hence, generators of $N_*(BS^1)$, ($S^1 \cong U_1$), are of the form $S(V_1^{2n+1})$, $n \geq 0$ while generators of $N_*(B(S^1/(\mathbb{Z}/m)))$ are $S(V_m^{2n+1})$, $n \geq 0$. It is however convenient to use a different set of generators. These arise by writing BS^1 as $B(S^1 \times_{\mathbb{Z}/2} \mathbb{Z}/2)$ and $B(S^1/(\mathbb{Z}/m))$ as $B(S^1 \times_{\mathbb{Z}/2m} \mathbb{Z}/2)$.

4.8.5 Generators of $N_*(B(S^1/(\mathbb{Z}/m)))$ are

$$S^1 \times_{\mathbb{Z}/2m} S^{2n}$$

for $n \geq 0$, where the generator of $\mathbb{Z}/2m$ acts on S^{2n} by the antipodal map.

It is a simple calculation, which we leave to the reader, to show that the set of $S^1 \times_{\mathbb{Z}/2m} S^{2n}$, $n \geq 0$, do generate

$N_*(B(S^1/(\mathbb{Z}/m)))$.

4.8.6 (i) $N_*^G[\mathbb{Z}/(2m+1);\Pi V_{k(i)}]$ is generated by

$$S^1 \times_{\mathbb{Z}/2(2m+1)} S^{2n} \times \Pi \eta_{j(i)} \otimes V_{k(i)}$$

where $n \geq 0$ and if $V_{k(q)} = V_{k(r)}$ for $q > r$ then $j(q) \geq j(r)$.

(ii) $N_*^G[S^1;\Pi V_{k(i)}]$ is generated by

$$\Pi \eta_{j(i)} \otimes V_{k(i)}$$

where if $V_{k(q)} = V_{k(r)}$ for $q > r$ then $j(q) \geq j(r)$.

(iii) $N_*^G[\mathbb{Z}/2m;\widetilde{R}^{2j+1} \times \Pi V_{k(i)}]$ is generated by

$$S^1 \times_{\mathbb{Z}/2m} (\xi_I \otimes \widetilde{R}) \times \Pi \eta_{j(i)} \otimes V_{k(i)}$$

where I is a sequence of non-increasing non-negative integers of length $2j+1$. Furthermore, if $j = 0$ then I (of length 1) is an even integer.

(iv) $N_*^G[\mathbb{Z}/2m;\widetilde{R}^{2j} \times \Pi V_{k(i)}]$ is generated by

$$S^1 \times_{\mathbb{Z}/2m} (\xi_I \otimes \widetilde{R}) \times \Pi \eta_{j(i)} \otimes V_{k(i)}$$

$$S^1 \times_{\mathbb{Z}/4m} S^{2n} \times (\eta_{2J} \otimes V_m) \times \Pi \eta_{j(i)} \otimes V_{k(i)}$$

where I, J are sequences of non-increasing non-negative integers of length $2j$ and j respectively.

To prove 4.8.6(iv) we need to know the cohomology of $B(S^1 \times_{\mathbb{Z}/2} O_{2j})$. We calculate this as in section 3.3 by looking at the Serre spectral sequence of the fibration

$$BO_{2j} \to B(S^1 \times_{\mathbb{Z}/2} O_{2j}) \to B(S^1/(\mathbb{Z}/2)).$$

It is not difficult, following section 3.3 to calculate the cohomology of $B(S^1 \times_{\mathbb{Z}/2} O_{2j})$.

4.8.7 $H^*(B(S^1 \times_{\mathbb{Z}/2} O_{2j}))$ is the $\mathbb{Z}/2[\beta, u_1, u_3, \ldots, u_{2j-1}, v_4, v_8, \ldots, v_{4j}]$ module on ρ_S, $S \in S_{2n}$ with relations $\beta u_{2i-1} = \beta v_{4i} = 0$. The u_{2i-1}, v_{4i}, ρ_S are as defined in 3.3.3 while β is the element of $H^2(B(S^1 \times_{\mathbb{Z}/2} O_{2j}))$ which comes from $\beta \in H^2(BS^1)$ where $H^*(BS^1) \cong \mathbb{Z}/2[\beta]$.

It is then not difficult to show that $N_*(B(S^1 \times_{\mathbb{Z}/2} O_{2j}))$ is generated by

$$S^1 \times_{\mathbb{Z}/2} (\xi_I \otimes \tilde{R})$$

$$S^1 \times_{\mathbb{Z}/4} S^{2n} \times (\eta_{2J} \otimes V_1) \quad .$$

To continue the proof of the theorem we need to define families F_j ($j \geq 0$) of S^1 slice types and induct over these families. The definitions of the families follows the general procedure described in the last section. We then define subsets A_j, B_j of F_j inductively. Let

$$A_0 = \{[1,0]\}, \quad B_0 = \emptyset .$$

Suppose that A_{j-1} and B_{j-1} are defined ($j \geq 1$). Let

$\{\rho_j\} = F_j - F_{j-1}$. If ρ_j is of the form

$$\rho_j = [\mathbb{Z}/2m; \tilde{R} \times \Pi V_{k(i)}]$$

with $1 \leq k(i) \leq m/2$ then set

$$A_j = A_{j-1} - \{[\mathbb{Z}/m; \Pi V_{k(i)}]\}$$
$$B_j = B_{j-1} \cup \{\rho_j\} \ .$$

Also, denote $[\mathbb{Z}/m; \Pi V_{k(i)}]$ by $r(\rho_j)$. If ρ_j is not of the above form then we set $A_j = A_{j-1} \cup \{\rho_j\}$ and $B_j = B_{j-1}$. As usual

$$A_j \cup B_j \cup r(B_j) = F_j \ .$$

4.8.8 If $G = S^1$ and $j \geq 0$ then $N_n^G[F_j]$ is isomorphic to the direct sum of

$$\bigoplus_{\rho \in A_j} N_n^G[\rho]$$

$$\bigoplus_{\rho \in B_j} (ker(\nu \partial : N_n^G[\rho] \to N_{n-1}^G[r(\rho)]))$$

$$\bigoplus_{\rho \in B_j} N_n^G[r(\rho)]/(im(\nu \partial : N_{n+1}^G[\rho] \to N_n^G[r(\rho)])) .$$

Furthermore, the isomorphism from the direct sum to $N_n^G[F_j]$, when restricted to the first and last terms is given by some (obvious) projective space construction.

Notice that as j gets large A_j consists of

$$[\mathbb{Z}/2m; \tilde{R}^{2k+1} \times \Pi V_{k(i)}] \ , \quad k > 1$$

$$[\mathbb{Z}/2m; \tilde{R} \times \Pi V_{k(i)}] \quad \text{with some } k(i) \text{ satisfying } m/2 < k(i) < m$$

$$[S^1; \Pi V_{k(i)}]$$

while B_j consists of

$$[\mathbb{Z}/2m; \tilde{R} \times \Pi V_{k(i)}] \qquad 1 \leq k(i) \leq m/2.$$

Finally, $r(B_j)$ consists of

$$r[\mathbb{Z}/2(2m+1); \tilde{R} \times \Pi V_{k(i)}] = [\mathbb{Z}/(2m+1); \Pi V_{k(i)}] \text{ for } m \geq 0,$$

$1 \leq k(i) \leq m$, and

$$r[\mathbb{Z}/4m; \tilde{R} \times \Pi V_{k(i)} \times V_m^k] = [\mathbb{Z}/2m; \tilde{R}^{2k} \times \Pi V_{k(i)}]$$

for $m \geq 1$, $1 \leq k(i) < m$.

Suppose now that $\rho \in B_j$ and suppose that ρ is not of the form

$$[\mathbb{Z}/4m; \tilde{R} \times \Pi V_{k(i)} \times V_m^k]$$

where $1 \leq k(i) < m$ and $k > 0$. In this case

$$\nu\partial : N_*^G[\rho] \to N_*^G[r(\rho)]$$

is an isomorphism. This can be proved by looking at the generators of each or by the fact that if

$$\rho = [\mathbb{Z}/2m; \tilde{R} \times \prod_{1 \leq i \leq (m-1)/2} V_i^{d(i)}]$$

then

$$N_*^G[\rho] \cong N_{*-1-\Sigma d(i)}(B((S^1/(\mathbb{Z}/m)) \times_{\mathbb{Z}/2} O_1)) \times \Pi BU_{d(i)})$$

$$\cong N_{*-1-\Sigma d(i)}(B(S^1/(\mathbb{Z}/m)) \times \Pi BU_{d(i)})$$

$$\cong N_*^G[r(\rho)] .$$

We are left with the case

$$\rho = [\mathbb{Z}/4m; \tilde{R} \times \Pi V_{k(i)} \times V_m^k]$$
$$r(\rho) = [\mathbb{Z}/2m; \tilde{R}^{2k} \times \Pi V_{k(i)}] .$$

The calculation of the kernel and image of

$$\nu\partial : N^G_*[\rho] \to N^G_{*-1}[r(\rho)]$$

is much the same as the calculations presented in sections 4.3 and 4.4. Indeed, the kernel of $\nu\partial$ is the N_* submodule of $N^G_*[\rho]$ generated by

$$S^1 \times_{\mathbb{Z}/4m} (\xi_{2a+2} \otimes \tilde{R}) \times (\eta_{2b} \otimes V_m) \times (\eta_I \otimes V_m) \times \Pi(\eta_{j(i)} \otimes V_{k(i)})$$

$$+ S^1 \times_{\mathbb{Z}/4m} (\xi_{2a+2} \otimes \tilde{R}) \times ((\eta_{2b} \otimes V_m) \times (\xi_{2a+2} \otimes \tilde{R})) \times (\eta_I \otimes V_m) \times \Pi(\eta_{j(i)} \otimes V_{k(i)})$$

where $a, b \geq 0$, $I = (i_2, i_3, \ldots, i_k)$ with $i_1 \geq i_2 \geq \ldots \geq i_k$ and $2b+1 \geq \max\{i_j; i_j \text{ odd}\}$. In fact these together with

$$S^1 \times_{\mathbb{Z}/4m} (\xi_{2a} \otimes \tilde{R}) \times (\eta_{2J} \otimes V_m) \times \Pi(\eta_{j(i)} \otimes V_{k(i)})$$

form an N_* basis for $N^G_*[\rho]$.

The remainder of the proof of the theorem is similiar to the proofs given in sections 4.3 and 4.4.

4.8.9 EXERCISE. Let F be a family of subgroups of $G = S^1$. Write down the generators of $N^G_*(F)$.

4.9 HISTORICAL NOTE

The first non-trivial group G for which generators of N^G_* were given was for $G = \mathbb{Z}/2$ in [Alexander] and in an unpublished manuscript of R.E. Stong. The generators given there are different to those given here.

The algebraic results concerning N^G_* for $G = (\mathbb{Z}/2)^k$ have been obtained by R.E. Stong (using slightly different methods) in [Stong, 3,4]. The author is indebted to J.S. Rose for proving the group theoretic results of section 4.5 (4.5.12, 4.5.13 etc).

R.P. Beem has recently also found generators of N_*^G for G odd or for G with sylow 2-subgroup $\mathbb{Z}/2$ in [Beem, 2].

The reader may like to look at some related results for generators of the oriented and unitary $\mathbb{Z}/p$ bordism rings in [Kosniowski, 2,3,4,5]. Rational generators of the oriented G bordism rings may be found in [Ossa]. A forthcoming paper by the author and E. Ossa will describe generators of the oriented $\mathbb{Z}/2$ bordism ring.

5 Generators of the equivariant cutting and pasting groups

5.1 A GENERAL REMARK

There are two interrelated methods of calculating the generators of the equivariant SK groups. The first reduces to the following steps.

(i) Prove exactness of the sequence

$$0 \to SK^G_*[F'] \xrightarrow{i} SK^G_*[F] \xrightarrow{\nu} SK^G_*[H;U]$$

for suitable families $F' \subseteq F$ with $F-F' = \{[H;U]\}$. We know that i is injective and $\nu i = 0$. So we need only show that ker $\nu \subseteq$ im i. In certain cases we have already done this in chapter 2. This step is by far the most difficult, in general.

(ii) Prove that $\nu(M) = k\nu'(M)$ for all $M \in SK^G_*[F]$, where k is some integer (often 1) and ν' is some homomorphism

$$\nu': SK^G_*[F] \to SK^G_*[H;U] \quad .$$

(iii) Prove that the sequence

$$0 \to SK^G_*[F'] \xrightarrow{i} SK^G_*[F] \xrightarrow{\nu'} SK_*[H;U] \to 0$$

is a short exact sequence.

(iv) Find a splitting for the exact sequence in (iii) and use the fact that $SK^G_*[H;U] \cong SK_*$,

$$SK_* \xrightarrow{\cong} SK^G_*[H;U]$$

$$M \longmapsto M \times (G \times_H U)$$

to find the generators of SK^G_* inductively.

196

The second (related) method is to use Theorem 2.7.1 and knowledge of the bordism splittings of

$$N_*^G[F] \xrightarrow{\nu} N_*^G[H;U]$$

i.e., the homomorphisms from im ν to $N_*^G[F]$. The procedure is to show that if M is a G manifold of type F then there is a G manifold L such that

(a) $\nu(M) \sim \nu(L)$ in $N_*^G[H;U]$, and

(b) $[L] = [L'][F(H;U)]$

where L' has trivial G action and $F(H;U)$ is a G manifold defined in terms of the slice type $[H;U]$. (The manifold L is zero if $\nu:N_*^G[F] \to N_*^G[H;U]$ is the zero homomorphism. In this case we let $F(H;U)$ be $G \times_H S(R \times U)$.) From Theorem 2.7.1 we obtain

$$[M] = [L'][F(H;U)] + [M'] - [M''] + \alpha[N][G \times_H S(R \times U)]$$

where M' and M'' are G manifolds of type F' while N has trivial G action and $\alpha = \pm 1$. We then show that

$$[G \times_H S(R \times U)] = k[F(H;U)]$$

and by induction deduce that the set of $F(H;U)$ with $[H;U] \in F$ generate $SK_*^G[F]$ as an SK_* module.

The two methods above are related in that the second method is used, in many cases, for steps (i) and (ii) of the first method.

5.2 GENERATORS OF SK_*^G; G OF ODD ORDER

Most of the steps mentioned in 5.1 (method 1) have already been completed. Indeed, Corollary 2.6.4 and Theorem 3.5.1 tell us that

$$SK_*^G \cong \bigoplus_{[H;U] \varepsilon St(G)} SK_*.$$

The result we obtain is

5.2.1 THEOREM. *If G is a finite abelian group of odd order then SK_*^G is a free SK_* module with basis consisting of the set of G manifolds of the form $G \times_H RP(R \times V)$ where H is a subgroup of G and V is an H module with no trivial H submodules, i.e. $V^H = \{0\}$.*

The proof is immediate from precise knowledge of the isomorphism $\oplus SK_* \to SK_*^G$.

5.2.2 REMARK. Instead of $G \times_H RP(R \times V)$ we could use
$$G/H \times \prod_{i=1}^{k} RP(R \times V_i)$$
where $V = V_1 \times V_2 \times \ldots \times V_k$.

5.2.3 COROLLARY. *If G is a finite abelian group of odd order then SK_*^G is multiplicatively generated as an SK_* module by*

$\{G/H; H \subseteq G\} \cup$

$\{RP(R \times V); V$ *is a non-trivial irreducible G module*$\}$.

If $[H;U]$ is a G slice type and M is a G manifold then denote by $M_{[H;U]}$ the set of points in M with slice type precisely $[H;U]$. We immediately have the following result.

5.2.4 COROLLARY. *Two n dimensional G manifolds M, N are equi-*

variantly SK equivalent (G finite abelian odd order) if and only if

$$\chi(M_{[H;U]}) = \chi(N_{[H;U]})$$

for all G slice types $[H;U] \in St(G)$.

Alternatively we could define $M^{[H;U]}$ to be the set of points in M with slice type contained in $G \times_H U$. The corollary would then require that

$$\chi(M^{[H;U]}) = \chi(M^{[H;U]})$$

for all G slice types $[H;U] \in St(G)$.

5.2.5 DEFINITION. A G-SK invariant is a homomorphism

$$SK_n^G \to \mathbb{Z}.$$

If $G = \{1\}$ then all SK invariants are multiples of the euler characteristic.

Define $\chi^{[H;U]}$ by

$$\chi^{[H;U]}(M) = \chi(M_{[H;U]}).$$

If V is a K module with $V^K = \{0\}$ then

$$\chi^{[H;U]}(G \times_K RP(R \times V)) = \begin{cases} |G/H| & \text{if } H = K, \ U = V \\ 0 & \text{otherwise} \end{cases}.$$

We therefore have the following generalisation to groups of odd order.

5.2.6 COROLLARY. If G is a finite abelian group of odd order then any G-SK invariant is a linear combination of the invariants

$$\{\chi^{[H;U]}/|G/H|\,;\,[H;U]\,\in\,St(G)\}\,.$$

5.3. GENERATORS OF $SK_*^{\mathbb{Z}/2}$.

The main result to be proved in this section is the following

5.3.1 THEOREM. $SK_*^{\mathbb{Z}/2}$ *is a free* SK_* *module with basis*

$$\{\mathbb{Z}/2\} \cup \{RP(R \times \tilde{R}^j)\,;\, j \geq 0\}$$

where $\tilde{R}$ *denotes the real numbers with* $\mathbb{Z}/2$ *acting by multiplication by* -1.

The proof is by induction on the families of $\mathbb{Z}/2$ slice types:

$$F_j = \{\sigma_{-1}, \sigma_0, \ldots, \sigma_j\}\,,\quad j \geq -1$$

where $\sigma_{-1} = [1;0]$ and $\sigma_j = [\mathbb{Z}/2; \tilde{R}^j]$ for $j \geq 0$. To start the induction we have

$$SK_*^G[\sigma_{-1}] \cong SK_*(B\mathbb{Z}/2) \cong SK_*\,.$$

Thus $SK_*^G[F_{-1}] = SK_*^G[\sigma_{-1}]$ is a free SK_* module with basis $\mathbb{Z}/2$.

We have the following sequence

$$0 \to SK_*^G[F_{j-1}] \xrightarrow{i} SK_*^G[F_j] \xrightarrow{\nu} SK_*^G[\sigma_j]$$

which we will show to be exact. If $j \neq 1$ then this follows from Corollary 2.6.2. In fact, if $j \neq 1$ then there is a short split exact sequence

$$0 \to SK_*^G[F_{j-1}] \to SK_*^G[F_j] \to SK_*^G[\sigma_j] \to 0$$

with a splitting

$$q: SK_*^G[\sigma_j] \to SK_*^G[F_j]$$

given by $q(E) = RP(R \times E)$.

Consider now the case $j = 1$.

5.3.2 LEMMA. *The sequence*
$$0 \to SK_n^G[F_0] \xrightarrow{i} SK_n^G[F_1] \xrightarrow{\nu} SK_n^G[\sigma_1]$$
is exact.

PROOF. Let $[M_1] - [M_2] \in \ker \nu$, so $[\nu(M_1)] = [\nu(M_2)]$ in $SK_n^G[\sigma_1]$. Then
$$\nu : N_n^G[F_1] \to N_n^G[\sigma_1]$$
is the zero homomorphism (see section 4.2) we have $\nu(M_1) \sim \nu(M_2)$. Hence by Corollary 2.7.2 we know that there are G manifolds M', M'' of type F_0 such that
$$[M_1] - [M_2] = [M'] - [M''].$$
Thus im $i = \ker \nu$ and the sequence is exact.

5.3.3 LEMMA. *There is a homomorphism ν'*
$$\nu' : SK_n^G[F_1] \to SK_n^G[\sigma_1]$$
such that $\nu = 2\nu'$.

PROOF. From Theorem 2.7.1 and the fact that
$$\nu : N_n^G[F_1] \to N_n^G[\sigma_1]$$
is the zero homomorphism we deduce that
$$[M] = [M'] - [M''] + \alpha[N][S(R \times \tilde{R})]$$
where M' and M'' are of type F_0 while N has trivial $\mathbb{Z}/2$ action

and $\alpha = \pm 1$. Thus

$$[\nu(M)] = 2\alpha[N][\tilde{R}]$$

which shows that $\nu = 2\nu'$ for some homomorphism ν'.

5.3.4 REMARK. The above lemma could be proved in the following way. Consider the sequence of homomorphisms

$$SK_n^G[F_1] \xrightarrow{\nu} SK_n^G[\sigma_1] \xrightarrow[\cong]{f} SK_{n-1} \xrightarrow[\cong]{\chi} \begin{cases} \mathbb{Z} & n \text{ odd} \\ 0 & n \text{ even} \end{cases}$$

where f denotes taking the fixed point set under $G = \mathbb{Z}/2$, or in other words the homomorphism which sends a bundle to its base space. Now, if M is a manifold of type F_1 then $\chi f \nu(M)$ is the euler characteristic of the codimension one fixed point set F in M. This number is even because if M' denotes the components of M with no points having slice type σ_0 then F is the boundary of the mapping cylinder of $M' \to M'/(\mathbb{Z}/2)$. Since χ and f are isomorphisms it follows that ν is divisible by 2, i.e., $\nu = 2\nu'$ for some ν'.

5.3.5 COROLLARY. *The following sequence*

$$0 \to SK_n^G[F_0] \xrightarrow{i} SK_n^G[F_1] \xrightarrow{\nu/2} SK_n^G[\sigma_1] \to 0$$

is a split short exact sequence.

PROOF. That the sequence is exact follows from the above two lemmas and the fact that $SK_n^G[\sigma_1]$ has no 2-torsion. To get a splitting define q

$$q: SK_n^G[\sigma_1] \to SK_n^G[F_1]$$

by $q(E) = RP(R \times E)$ for $E \in SK_n^G[\sigma_1]$. We see that

$$(\nu/2)q(E) = E$$

thereby establishing the result.

We conclude that

$$SK_*^{\mathbb{Z}/2} \cong \bigoplus_{j \geq -1} SK_*^{\mathbb{Z}/2}[\sigma_j].$$

We know from section 3.5 that $SK_*^G[\sigma_j]$, for $j \geq 0$, is a free SK_* module on the generator $\tilde{R}^j$. Thus Theorem 5.3.1 follows immediately.

5.3.6 REMARK. An alternative way of proving Theorem 5.3.1 would be by the second method mentioned in section 5.1. We illustrate this now. Suppose M is a G manifold of type F_j, $j \geq 0$, then there is a manifold

$$L = \sum_I N_I\, RP(R \times \xi_I)$$

where $I = (i_1, i_2, \ldots, i_j)$ with $i_1 \geq i_2 \geq \ldots \geq i_j \geq 0$, such that $\nu(M) \sim \nu(L)$ in $N_n^G[\sigma_j]$. (This statement is true for $j = 1$.) We have

$$[L] = [L'] [RP(R \times \tilde{R}^j)]$$

using Theorem 2.4.1. Hence by Theorem 2.7.1 we have

$$[M] = [L'][RP(R \times \tilde{R}^j)] + [M'] - [M''] + \alpha[N][S(R \times \tilde{R}^j)]$$

where M' and M'' are of type F_{j-1}. Since

$$[S(R \times R^j)] = k[RP(R \times R^j)]$$

where $k = 1$ if $j = 1$ and $k = 2$ if $j \neq 1$ we deduce that

$$[M] = [M'] - [M''] + ([L'] + k\alpha[N])[RP(R \times \tilde{R}^j)].$$

The required result then follows by induction.

5.3.7 COROLLARY. *Let M, M' be n dimensional $\mathbb{Z}/2$ manifolds and let $F_0, F_1, \ldots, F_n$, $(F'_0, F'_1, \ldots, F'_n)$ be the fixed point sets of $M(M')$ of codimensions $0, 1, \ldots, n$ respectively. Then M and M' are equivariantly SK equivalent if and only if $\chi(M) = \chi(M')$ and $\chi(F_i) = \chi(F'_i)$ for $i = 0, 1, \ldots, n$.*

PROOF. One way to prove this result is first to separate the two cases $n = 2m$, $n = 2m+1$ and then write a typical manifold in terms of those given in Theorem 5.3.1.

Case (i) $n = 2m$. If M is a $2m$ dimensional $\mathbb{Z}/2$ manifold then we may write $[M]$ as

$$\sum_{j=0}^{2m} \alpha_{2j} [(RP^2)^{m-j}][RP(R \times \tilde{R}^{2j})] + \alpha[(RP^2)^m][\mathbb{Z}/2] \ .$$

Thus

$$[F_i] = \begin{cases} \alpha_{2j} [(RP^2)^{m-j}] & \text{if } i = 2j \\ 0 & \text{if } i \text{ is odd and } i \neq 1 \\ \sum \alpha_{2j} [(RP^2)^{m-j}][RP^{2j-1}] & \text{if } i = 1 \end{cases}$$

and

$$\chi(M) = 2\alpha + \sum \alpha_{2j}$$

$$\chi(F_i) = \begin{cases} \alpha_{2j} & \text{if } i = 2j \\ 0 & \text{if } i \text{ is odd.} \end{cases}$$

Case (ii) $n = 2m+1$. We may write $[M^{2m+1}]$ as

$$\sum_{j=0}^{m} \alpha_{2j+1} [(RP^2)^{m-j}][RP(R \times \tilde{R}^{2j+1})] \ .$$

So that

$$[F_i] = \begin{cases} \alpha_{2j+1}[(RP^2)^{m-j}] & \text{if } i = 2j+1, i \neq 1 \\ 2\alpha_1[(RP^2)^m] + \sum_{j \neq 1} \alpha_{2j+1}[(RP^2)^{m-j}][RP^{2j}] & \text{if } i = 1 \\ 0 & \text{if } i \text{ is even} \end{cases}$$

and

$$\chi(F_i) = \begin{cases} \alpha_{2j+1} & \text{if } i = 2j+1, i \neq 1 \\ 2\alpha_1 + \sum_{j \neq 0} \alpha_{2j+1} & \text{if } i = 1 \\ 0 & \text{otherwise} \end{cases}.$$

The result follows immediately.

Having done the above calculations it is now easy to get the next result.

5.3.8 COROLLARY. *(i) Any $\mathbb{Z}/2$-SK invariant for $2n$ dimensional $\mathbb{Z}/2$ manifolds is a linear combination of*

$$\tfrac{1}{2}(\chi+\chi_0+\chi_2+\ldots+\chi_{2n}), \chi_0, \chi_2, \ldots, \chi_{2n}$$

where $\chi_i(M)$ is the euler characteristic of the codimension i fixed point set.

(ii) Any $\mathbb{Z}/2$-SK invariant for $(2n+1)$ dimensional $\mathbb{Z}/2$ manifolds is a linear combination of

$$\tfrac{1}{2}(\chi_1+\chi_3+\ldots+\chi_{2n+1}), \chi_3, \chi_5, \ldots, \chi_{2n+1}.$$

5.3.9 REMARK. (i) $\chi(M/(\mathbb{Z}/2)) = \tfrac{1}{2}(\chi(M)+\chi(M^{\mathbb{Z}/2}))$
$= \tfrac{1}{2}(\chi+\chi_0+\chi_2+\ldots)(M).$

(ii) $(X_1+X_3+\ldots+X_{2n+1})(M^{2n+1}) = \chi(M^{\mathbb{Z}/2})$
$= \chi(M) \bmod 2$
$= 0 \bmod 2.$

An alternative, and better, way of proving the previous two corollaries is from the next result.

5.3.9 LEMMA. *There is an SK_* isomorphism*

$$SK_*^G \to SK_* \oplus \bigoplus_{j \geq 0} SK_*$$

PROOF. Define SK_* homomorphisms

$$f_i : SK_*^G \to SK_*$$

for $i = -1, 0, 1, 2, \ldots$ by

$$f_i[M] = \begin{cases} M + \sum_{j \geq 0} [(RP^2)^j][F_{2j}] & \text{if } i = -1 \\ \sum_{j \geq 0} [(RP^2)^j][F_{2j+1}] & \text{if } i = 1 \\ [F_i] & \text{if } i = 0, 2, 3, \ldots \end{cases}$$

It is easy to check that if M is a basis element of SK_*^G then $f_{-1}[M] = 2[X]$ and $f_1[M] = 2[Y]$ for some X and Y. Hence this also holds for all manifolds. We define an SK_* homomorphism

$$f : SK_*^G \to \bigoplus_{j \geq -1} SK_*$$

by

$$f[M] = (\tfrac{1}{2}f_{-1}[M], f_0[M], \tfrac{1}{2}f_1[M], f_2[M], f_3[M], \ldots).$$

By ordering the basis elements of SK_*^G as

$\mathbb{Z}/2, RP(R\times\tilde{R}^0), RP(R\times\tilde{R}^1), \ldots$

it is easy to see that the matrix of f is triangular with one's down the diagonal. Hence f is an isomorphism.

Corollary 5.3.7, for example, follows from this lemma simply by taking euler characteristics

5.3.10 REMARK. We could use the following homomorphisms instead of the f_i given above.

$$[M] \mapsto \begin{cases} [M] - \sum_{j\geq 0} [(RP^2)^j][F_{2j}] & \text{if } i = -1 \\ [F_1] - \sum_{j\geq 1} [(RP^2)^j][F_{2j+1}] & \text{if } i = 1 \\ [F_i] & \text{if } i = 0, 2, 3, \ldots \end{cases}$$

5.4 GENERATORS OF $SK_*^{\mathbb{Z}/4}$.

Let $\tilde{R}$ denote the real numbers with $\mathbb{Z}/4$ and $\mathbb{Z}/2$ acting by multiplication by -1. Let $\tilde{C}$ denote the complex numbers with $\mathbb{Z}/4$ acting by multiplication by i.

5.4.1 THEOREM. $SK_*^{\mathbb{Z}/4}$ is a free SK_* module with basis

$\{\mathbb{Z}/4\} \cup$

$\{\mathbb{Z}/4 \times_{\mathbb{Z}/2} RP(R\times\tilde{R}^j); j \geq 0\} \cup$

$\{RP(R\times\tilde{R}^j) \times RP(R\times\tilde{C}^k); j, k \geq 0\}$.

Recall that the $\mathbb{Z}/4$ slice types are

$\sigma_{-1} = [1; 0]$

$\sigma_j = [\mathbb{Z}/2; \tilde{R}^j]$, $j \geq 0$

$$\sigma_{j,k} = [\mathbb{Z}/4; \tilde{R}^j \times \tilde{C}^k] \, , \quad j,k \geq 0 \, .$$

We order these as in section 4.3 and let F_j be the family

$$F_j = \{\rho_0, \rho_1, \ldots, \rho_j\}$$

where $\rho_0, \rho_1, \ldots$ are the ordered $\mathbb{Z}/4$ slice types. From Corollary 2.6.3 we immediately obtain the fact that the sequence

$$0 \to SK_*^G[F_{j-1}] \xrightarrow{i} SK_*^G[F_j] \xrightarrow{\nu} SK_*^G[\rho_j] \to 0$$

is a split short exact sequence if $\rho_j \neq \sigma_1$ or $\sigma_{1,l}$. A splitting q

$$q: SK_*^G[\rho_j] \to SK_*^G[F_j]$$

is given by $q(E) = RP(R \times E)$. In the case that $\rho_j = \sigma_{k,l}$ we may use an alternative splitting by writing E as a direct sum $E = E_0 \oplus E_1$, where E_0 corresponds to the subbundle with action determined by $\tilde{R}$ and E_1 corresponds to the subbundle with action determined by $\tilde{C}$. In fact E_0 is a k dimensional vector bundle while E_1 is a $2l$ dimensional vector bundle. We can define a splitting by sending E to $RP(R \times (E_0, E_1))$ where $RP(R \times (E_0, E_1))$ is defined in the proof of Lemma 4.4.2. It remains to deal with the cases $\rho_j = \sigma_1$ or $\sigma_{1,l}$.

5.4.2 THEOREM. *The sequence*

$$0 \to SK_n^G[F_{j-1}] \xrightarrow{i} SK_n^G[F_j] \xrightarrow{\nu} SK_n^G[\rho_j]$$

is an exact sequence.

PROOF. Of course we only have to give a proof in the cases $\rho_j = \sigma_1$ or $\sigma_{1,l}$. If $\rho_j = \sigma_1$ or $\sigma_{1,0}$ then the result follows

from Corollary 2.7.2 and the fact that
$$\nu: N_*^G[F_j] \to N_*^G[\rho_j]$$
is the zero homomorphism.

For the case $\rho_j = \sigma_{1,l}$ ($l > 0$) we first prove the following result.

5.4.3 LEMMA. *If M is a $\mathbb{Z}/4$ manifold of type F_j, where $\rho_j = \sigma_{1,l}$ with $l \geq 0$, then there are manifolds M_1, M_2 of type F_{j-1} and manifolds B_1, B_2 with trivial $\mathbb{Z}/4$ action such that*
$$[M] = [M_1] - [M_2] + ([B_1] - [B_2])[RP(R \times \tilde{R})][RP(R \times C^l)].$$

PROOF. From section 4.3 we know that there is a $\mathbb{Z}/4$ manifold L of the form
$$L = \sum \alpha_{a,b,J} \, RP(R \times \xi_{a+2} \times \eta_{2b}) \times RP(R \times \eta_J)$$
where $\alpha_{a,b,J}$ are manifolds with trivial $\mathbb{Z}/4$ action, such that $\tilde{\nu}(M) \sim \nu(L)$ in $N_n^G[\rho_j]$. (We put $L = 0$ if $l = 0$.) From Theorem 2.4.1 we have
$$[L] = \sum [\alpha_{a,b,J}][RP^{a+2} \times CP^{2b} \times CP^J][RP(R \times \tilde{R} \times \tilde{C})][RP(R \times \tilde{C}^{l-1})].$$

We have the following bundle equivalences
$$\nu(S(R \times \tilde{R} \times \tilde{C}^l)) = \nu(RP(R \times \tilde{R} \times \tilde{C}) \times RP(R \times \tilde{C}^{l-1}))$$
$$= \nu(RP(R \times \tilde{R}) \times RP(R \times \tilde{C}^l)).$$

From 2.6.1 we therefore have that
$$S(R \times \tilde{R} \times \tilde{C}^l),$$
$$RP(R \times \tilde{R} \times \tilde{C}) \times RP(R \times \tilde{C}^{l-1})$$

and

$$RP(R \times \tilde{R}) \times RP(R \times \tilde{C}{}^l)$$

are all $SK^{\mathbb{Z}/4}$ equivalent modulo $\mathbb{Z}/4$ manifolds of type F_{j-1}. The result follows from Theorem 2.7.1.

To prove Theorem 5.4.2 in the case $\rho_j = \sigma_{1,l}$ ($l > 0$) suppose that $[M] - [M_0] \in SK_n^G[F_j]$ with $[\nu(M)] = [\nu(M_0)]$. From 5.4.3 we see that

$$[M] - [M_0] = [M'] - [M''] + ([B'] - [B''])[RP(R \times \tilde{R})][RP(R \times \tilde{C}{}^l)]$$

where M' and M'' are of type F_{j-1} while B' and B'' have trivial $\mathbb{Z}/4$ action. Since $[\nu(M)] = [\nu(M_0)]$ we see that

$$2([B'] - [B''])[\tilde{R} \times \tilde{C}{}^l] = 0 \text{ in } SK_n^G[\sigma_{1,l}]$$

which means that $[B'] - [B''] = 0$ since $SK_n^G[\sigma_{1,l}]$ is isomorphic to SK_{n-1-2l} which has no 2-torsion. This completes the proof of Theorem 5.4.2.

The next stage in the proof of Theorem 5.4.1 is to show that there is always a short exact sequence

$$0 \to SK_*^G[F_{j-1}] \overset{i}{\to} SK_*^G[F_j] \overset{\nu'}{\to} SK_*^G[\rho_j] \to 0$$

Of course if $\rho_j \neq \sigma_1$ or $\sigma_{1,l}$ ($l > 0$) then we know that this is so with $\nu' = \nu$. If $\rho_j = \sigma_{1,l}$ with $l > 0$ then from Theorem 5.4.2 and Lemma 5.4.3 we see that there is such an exact sequence with $2\nu' = \nu$. Finally if $\rho_j = \sigma_1$ then there is such an exact sequence with $2\nu' = \nu$, this follows from the next lemma whose proof is immediate.

5.4.4 LEMMA. *If M is a $\mathbb{Z}/4$ manifold of type F_j, where $\rho_j = \sigma_1$ then there are manifolds M_1, M_2 of type F_{j-1} and manifolds B_1, B_2*

with trivial $\mathbb{Z}/4$ action such that

$$[M] = [M_1] - [M_2] + ([B_1] - [B_2])[\mathbb{Z}/4 \times_{\mathbb{Z}/2} RP(R \times \tilde{R})].$$

Finally, the homomorphisms

$$q_j : SK^G_*[\rho_j] \to SK^G_*[F_j]$$

defined by

$$q_j(E) = RP(R \times E) \qquad \text{if } \rho_j = \sigma_k$$
$$q_j(E = E_0 \oplus E_1) = RP(R \times (E_0, E_1)) \qquad \text{if } \rho_j = \sigma_{k,l}$$

are splitting homomorphisms for the exact sequence

$$0 \to SK^G_*[F_{j-1}] \xrightarrow{i} SK^G_*[F_j] \xrightarrow{\nu'} SK^G_*[\rho_j] \to 0.$$

The main result, Theorem 5.4.1, follows immediately.

We now turn our attention to the $\mathbb{Z}/4$-SK invariants. Consider the following SK_* homomorphisms

$$f_k, f_{k,l} : SK^{\mathbb{Z}/4}_* \to SK_*$$

where

$$f_k[M] = \begin{cases} [M] & \text{if } k = -1 \\ [F_k] & \text{if } k \geq 0 \end{cases}$$

and

$$f_{k,l}[M] = [F_{k,l}] \qquad k \geq 0, \, l \geq 0.$$

Here $F_{k,l}$ denotes the submanifold of M consisting of those points in M with slice type $\sigma_{k,l}$. The manifold F_{2k+1} is defined as the set of points in M with slice type σ_{2k+1} while F_{2k} is the set of points in M with slice type σ_{2k} or $\sigma_{j,k}$ (any $j \geq 0$). Let

M_i, $i \geq 0$ denote the manifold

$$M_i = \mathbb{Z}/4 \times_{\mathbb{Z}/2} RP(R \times \tilde{R}^i)$$

and let $M_{i,j}$ be given by

$$M_{i,j} = RP(R \times \tilde{R}^i) \times RP(R \times \tilde{C}^j).$$

The evaluation of f_k, $f_{k,l}$ on the manifolds $\mathbb{Z}/4, M_i, M_{i,j}$ is given in Table 5.4.5 (page 213).

From Table 5.4.5 we see that the following are also homomorphisms

$$g_{-1} = (\tfrac{1}{4})(f_{-1} - \sum_{k \geq 0} [RP^{2k}] f_{2k} - \sum_{k,l \geq 0} [RP^{2k}][RP^{2l}] f_{2k,l})$$

$$g_1 = (\tfrac{1}{4})(f_1 - \sum_{k \geq 1} [RP^{2k}] f_{2k+1})$$

$$g_{2k} = (\tfrac{1}{2})(f_{2k} - \sum_{i \geq 0} [RP^{2i}] f_{2i,k})$$

$$g_{2k+1} = (\tfrac{1}{2}) f_{2k+1} \quad (k \neq 0)$$

$$g_{1,l} = (\tfrac{1}{2})(f_{1,l} - \sum_{k \geq 1} [RP^{2k}] f_{2k+1,l})$$

$$g_{k,l} = f_{k,l} \quad (k \neq 1).$$

	$\mathbb{Z}/4$	M_1	M_{2i}	M_{2i+1} $(i>0)$	$M_{1,j}$	$M_{2i,j}$	$M_{2i+1,j}$ $(i>0)$
f_{-1}	$4[pt]$	0	$2[RP^{2i}]$	0	0	$[RP^{2i}][RP^{2j}]$	0
f_1	0	$4[pt]$	0	$2[RP^{2i}]$	0	0	0
f_{2k}	0	0	$2[pt]$ if $k=i$, 0 if $k \neq i$	0	0	$[RP^{2i}]$ if $k=j$, 0 if $k \neq j$	0
f_{2k+1} $(k>0)$	0	0	0	$2[pt]$ if $k=i$, 0 if $k \neq i$	0	0	0
$f_{1,l}$	0	0	0	0	$2[pt]$ if $l=j$, 0 if $l \neq j$	0	$[RP^{2i}]$ if $l=j$, 0 if $l \neq j$
$f_{2k,l}$	0	0	0	0	0	$[pt]$ if $k=i$, $l=j$; 0 otherwise	0
$f_{2k+1,l}$ $(k>0)$	0	0	0	0	0	0	$[pt]$ if $k=i$, $l=j$; 0 otherwise

5.4.5 TABLE

The homomorphism

$$g: SK_*^{\mathbb{Z}/4} \to \bigoplus_{i \geq -1} SK_* \oplus \bigoplus_{j,k \geq 0} SK_*$$

defined by $g = (\oplus g_i) \oplus (\oplus g_{j,k})$ is clearly seen to be an isomorphism. The next result follows immediately.

5.4.6 THEOREM. *Any equivariant SK invariant for $\mathbb{Z}/4$ manifolds is a linear combination of the following set of invariants:*

$$\{(\tfrac{1}{4})(X - \sum_{k \geq 0} X_{2k} - \sum_{k,l \geq 0} X_{2k,l})\} \cup$$

$$\{(\tfrac{1}{4})(X_1 - \sum_{k \geq 1} X_{2k+1})\} \cup$$

$$\{(\tfrac{1}{2})(X_{2k} - \sum_{i \geq 0} X_{2i,k}); k \geq 0\} \cup$$

$$\{(\tfrac{1}{2})X_{2k+1}; k \geq 1\} \cup$$

$$\{(\tfrac{1}{2})(X_{1,l} - \sum_{k \geq 1} X_{2k+1,l}); l \geq 0\} \cup$$

$$\{X_{k,l}; k,l \geq 0, k \neq 1\}$$

where $X_i, X_{i,j}$ denotes the euler characteristic of the points in a manifold with slice type contained in σ_i, $\sigma_{i,j}$ respectively.

5.4.7 COROLLARY. *Two n dimensional $\mathbb{Z}/4$ manifolds M, M' are equivariantly SK equivalent if and only if*

$$\chi(M) = \chi(M')$$
$$\chi_i(M) = \chi_i(M') \quad i = 0, 1, \ldots, n$$
$$\chi_{i,j}(M) = \chi_{i,j}(M') \quad i, j \geq 0, \ i + 2j \leq n.$$

5.4.8 REMARK. The following are trivial observations:

$$\left(\chi + \sum_{i \geq 0} \chi_i + 2 \sum_{i,j \geq 0} \chi_{i,j}\right)(M) = 0 \mod 4$$

$$\sum_{i \geq 0} \chi_i(M) = \chi(M^{\mathbb{Z}/2})$$

$$\sum_{i,j \geq 0} \chi_{i,j}(M) = \chi(M^{\mathbb{Z}/4})$$

$$\chi(M) + \chi(M^{\mathbb{Z}/2}) + 2\chi(M^{\mathbb{Z}/4}) = 0 \mod 4$$

in fact the exact value of the last expression is $4\chi(M/(\mathbb{Z}/4))$.

5.5 GENERATORS OF SK_*^G; $G = \mathbb{Z}/2^r$.

The main result of this section generalises the main result of the previous two sections.

5.5.1 THEOREM.. *A free SK_* base for SK_*^G, $G = \mathbb{Z}/2^r$ is given by the following $\mathbb{Z}/2^r$ manifolds*

(i) $\mathbb{Z}/2^r$

(ii) $\mathbb{Z}/2^r \times_{\mathbb{Z}/2^s} RP(R \times V_0^{i(0)}) \times \prod_{j=1}^{2^{s-1}-1} RP(R \times V_j^{i(j)})$

where $r \geq s > 0$, $i(j) \geq 0$ for $j = 0, 1, \ldots, 2^{s-1}-1$, and if $i(0) = 1$ then $i(2^{s-2}+1) = i(2^{s-2}+2) = \ldots = i(2^{s-1}-1) = 0$.

(iii) $\mathbb{Z}/2^r \times_{\mathbb{Z}/2^s} RP(R \times V_0 \times V_l^{i(l)}) \times \prod_{j=1}^{l-1} RP(R \times V_j^{i(j)})$

where $r \geq s > 2$, $i(j) \geq 0$ for $j = 0, 1, \ldots, l$ and $2^{s-2} < l < 2^{s-1}$.

By V_0 we mean the real numbers with $\mathbb{Z}/2^s$ acting as $\{\pm 1\}$ while by V_j ($j > 0$) we mean the complex numbers with a generator of $\mathbb{Z}/2^s$ (or of $\mathbb{Z}/2^r$) acting by multiplication by $\exp(2\pi i j/2^s)$ (or by $\exp(2\pi i j/2^r)$ respectively).

The proof of the theorem and its corollary will be left to the reader. The most difficult part of the proof has already been done in the previous section.

5.5.2 COROLLARY. *Two n dimensional $\mathbb{Z}/2^r$ manifolds, M and M', are equivariantly SK equivalent if and only if*

$\chi(M) = \chi(M')$

$\chi_{(A,s)}(M) = \chi_{(A,s)}(M')$ *for all s*, $0 < s \leq r$ *and all* $(2^{s-1}-1)$ *tuples A,*

where $\chi_{(A,s)}(M)$ *denotes* $\chi(M[H;U])$ *with*

$H = \mathbb{Z}/2^s$, $U = \prod_{j=1}^{2^{s-1}-1} V_j^{a(j)}$

if $A = (a_0, a_1, \ldots)$.

5.6 GENERATORS FOR SK_*^G; $G = (\mathbb{Z}/2)^k$.

Consider the case $G = (\mathbb{Z}/2)^2$ with generators g_1, g_2. The nontrivial irreducible G modules are V_1, V_2, V_3 where $V_i = V(G_i)$ for $i = 1, 2, 3$ and $G_1 = \langle g_1 \rangle$, $G_2 = \langle g_2 \rangle$, $G_3 = \langle g_1 + g_2 \rangle$. The G slice types are

$[0; 0]$

$[G_i; \tilde{R}^k]$ $i = 1, 2, 3$, $k \geq 0$

$[G; V_1^a \times V_2^b \times V_3^c]$ $a, b, c \geq 0$.

5.6.1 THEOREM. *If $G = (\mathbb{Z}/2)^2$ then SK_*^G is a free SK_* module with basis*

$G \times_{G_1} RP(R \times \tilde{R}^k)$ with $k \geq 0$,

$S(V_2 \times V_1^k)$ with $k \geq 0$,

$S(V_3 \times V_1^k)$ with $k \geq 0$,

$RP(R \times V_1^a) \times RP(R \times V_2^b \times V_3^c)$ with $a \neq 1$ and $b+c \neq 1$,

$S(R \times V_1) \times_{\mathbb{Z}/2} RP(R \times V_2^b \times V_3^c)$ with $c \neq 0$ and $b+c \neq 1$,

$S(R \times V_1^a \times V_2^b \times V_3^c)$ with $a = 1$, $c = 0$ or $b+c = 1$.

The action of $\mathbb{Z}/2$ on $S(R \times V_1) \times RP(R \times V_2^b \times V_3^c)$ is given by

$$(x_0, x_1) \times (y_1, y_2, y_3) \to (-x_0, -x_1) \times (y_1, y_2, -y_3)$$

where $(x_0, x_1) \in S(R \times V_1)$ and $(y_1, y_2, y_3) \in RP(R \times V_2^b \times V_3^c)$ with y_1, y_2, y_3 coming respectively from R, V_2^b, V_3^c.

To prove the result we define families F_j ($j \geq 0$) of G slice types as in section 4.5. We then have:

5.6.2 THEOREM. *The sequence*

$$0 \to SK_*^G[F_{j-1}] \to SK_*^G[F_j] \to SK_*^G[\rho_j]$$

is an exact sequence for $j \geq 0$ *and* $G = (\mathbb{Z}/2)^2$.

PROOF. Use the results of section 2.7 and Remark 4.5.21. There are two cases depending on whether the homomorphism

$$\nu: N_*^G[F_j] \to N_*^G[\rho_j]$$

is (i) zero or (ii) non-zero. In case (i) we use 2.7.2. For case (ii) we state some definitions and results.

5.6.3 DEFINITION. If $[H;U]$ is a G slice type then define $F(H;U)$ by

$$F(0;0) = G,$$
$$F(G_1; \tilde{R}^k) = G \times_{G_1} RP(R \times \tilde{R}^k)$$
$$F(G_2; \tilde{R}^k) = S(V_2 \times V_1^k)$$
$$F(G_3; \tilde{R}^k) = S(V_3 \times V_1^k)$$
$$F(G; V_1^a \times V_2^b \times V_3^c) = \begin{cases} RP(R \times V_1^a) \times RP(R \times V_2^b \times V_3^c) & \text{if } a \neq 1 \text{ and } b+c \neq 1 \\ S(R \times V_1) \times_{\mathbb{Z}/2} RP(R \times V_2^b \times V_3^c) & \text{if } c \neq 1 \text{ and } b+c \neq 1 \end{cases}$$

Although it is not necessary we may also define $F(G; V_1^a \times V_2^b \times V_3^c)$ in the case $a = 1$, $c = 0$ or $b+c = 1$ by $S(R \times V_1^a \times V_2^b \times V_3^c)$.

5.6.4 LEMMA. *Suppose that* $\rho_j = [H;U]$. *If M is a G manifold of type* F_j *then there is a G manifold L of type* F_j *such that*

(i) $\nu(M) \sim \nu(L)$, and

(ii) $[L] = [L'][F(H;U)]$

where L' has trivial G action.

PROOF. This follows easily from 4.5.21 and some easy calculations.

5.6.5 LEMMA. *If $[H;U]$ is a G slice type then*

$$\nu(G \times_H S(R \times U)) = m\nu(F(H;U))$$

where $m = 1$ if $[H;U] = [G_1; \tilde{R}]$ or $[H;U] = [G; V_1^a \times V_2^b \times V_3^c]$ with $a = 1$, $c = 0$, or $b+c = 1$, and otherwise $m = 2$.

PROOF. Easy calculations.

The proof of Theorem 5.6.2 then easily follows using Theorem 2.7.1 and Theorem 2.6.1.

5.6.6 COROLLARY. *The sequence*

$$0 \to SK_*^G[F_{j-1}] \overset{i}{\to} SK_*^G[F_j] \overset{\nu'}{\to} SK_*^G[\rho_j] \to 0$$

where $\nu' = (2/m)\nu$ with m given in 5.6.5, is a short split exact sequence.

We then easily obtain Theorem 5.6.1 and the following.

5.6.7 COROLLARY. *If $G = (\mathbb{Z}/2)^2$ then two n dimensional G manifolds M, N are G-SK equivalent if and only if*

$$\chi_\rho(M) = \chi_\rho(N)$$

for all G slice types ρ, where $\chi_\rho(M) = \chi(M^\rho)$.

The calculation of SK_*^G for $G = (\mathbb{Z}/2)^k$, $k \geq 3$, becomes complicated. We leave it for the interested reader.

5.7 GENERATORS FOR SK_*^G; G GENERAL

The main problem in general for calculating SK_*^G is to show that the sequence

$$0 \to SK_*^G[F'] \to SK_*^G[F] \to SK_*^G[\rho]$$

is exact for families F, F' of G slice types with $F - F' = \{\rho\}$. It seems reasonable to conjecture that this be so for finite abelian groups. The past few sections have shown that this is true for many groups. Another case for which the conjecture is true is groups of the form $G \times \mathbb{Z}/2$ where G is of odd order. We leave this as an exercise for the reader.

5.7.1 EXERCISE. Let G be a finite abelian group of odd order. Prove that $SK_*^{G \times \mathbb{Z}/2}$ is a free SK_* module with basis

$$(G/H) \times RP(R \times R^a) \times \prod_{i=1}^{k} RP(R \times V_i^{a(i)}) \times \prod_{i=1}^{k} RP(R \times V_i^{b(i)})$$

with $a \neq 1$ and $\Sigma b(i) \neq 0$ or $a = 1$ and $\Sigma b(i) = 0$,

$$(G/H) \times RP(R \times R \times V_j^{b(j)}) \times \prod_{i=1}^{k} RP(R \times V_i^{a(i)}) \times \prod_{i=j+1}^{k} RP(R \times V_i^{b(i)})$$

with $b(j) \neq 0$,

$$(G/H) \times (\mathbb{Z}/2) \times \prod_{i=1}^{k} RP(R \times V_i^{a(i)}) \quad .$$

Here H is a subgroup of G of order $2k+1$ and $V_1, V_2, \ldots, V_k$ are the non-trivial irreducible H modules (thought of as $G \times \mathbb{Z}/2$

modules) while $\tilde{V}_i = \tilde{R} \otimes V_i$.

Furthermore show that two $G \times \mathbb{Z}/2$ manifolds M, N are equivariantly SK equivalent if and only if

$$\chi_\rho(M) = \chi_\rho(N)$$

for all $G \times \mathbb{Z}/2$ slice types ρ.

It is also reasonable to conjecture in general that two manifolds M, N are G-SK equivalent if and only if

$$\chi_\rho(M) = \chi_\rho(N)$$

for all $\rho \in St(G)$. This is certainly true in $SK_*^G \otimes \mathbb{Z}[\tfrac{1}{2}]$ as can be seen quite easily - essentially from 2.7.3.

5.7.2 EXERCISE. Calculate SK_*^G for $G = S^1$.

5.7.3 PROBLEM. The Burnside Ring $\Omega(G)$ of a group G may be defined as the set of equivalence classes of G manifolds (no restriction on dimension) with the equivalence relation $M_1 \sim M_2$ if and only if $\chi(M_1^H) = \chi(M_2^H)$ for all subgroups H of G. We therefore have, in many cases, a homomorphism

$$SK_*^G \to \Omega(G)$$

What can you say about this homomorphism?

References

Adams, J.F.

"Lectures on Lie groups", New York, Benjamin 1969.

Alexander, J.C.

The bordism ring of manifolds with involutions. *Proc. Amer. Math. Soc.* 31 (1972), 536-542.

Atiyah, M.F.

"K-theory", New York, Benjamin 1967.

Beem, R.P.

[1] On the bordism of almost free $(\mathbb{Z}/2)^k$ actions. *Trans. Amer. Math. Soc.* 225 (1977) 83-105.

[2] Generators for G bordism. preprint.

Bierstone, E.

Equivariant Gromov Theory. *Topology* 13 (1974) 327-345.

Bredon, G.E.

"Introduction to compact transformation groups", New York and London, Academic Press, 1972.

Conner, P.E. and Floyd, E.E.

[1] "Differentiable periodic maps", Berlin-Heidelberg-New York Springer, 1964.

[2] Maps of odd period. *Ann. of Math.* 84 (1966), 132-156.

Curtis, C.W. and Reiner, I.

"Representation theory of finite groups and associative algebras", New York, Interscience 1962.

Dold, A.

[1] "Lectures on Algebraic Topology", Berlin-Heidelberg-New York, Springer 1972.

[2] Erzeugende der Thomschen Algebra N. *Math Z.* 65 (1956) 25-35.

Hilton, P.J. and Wylie, S.

"Homology Theory", Cambridge, Cambridge University Press 1960.

Hirsch, M.W.

"Differential Topology", New York-Heidelberg-Berlin, Springer 1976.

Husemoller, D.

"Fibre Bundles", New York-Heidelberg-Berlin, Springer, 1975 (second edition).

Janich, K.

[1] "Charakterisierung der Signatur von Mannigfaltigkeiten durch eine Additivitats-Eigenschaft. *Inventiones Math.* 6 (1968) 35-40.

[2] On invariants with the Novikov additivity property. *Math. Ann.* 184 (1969), 65-77.

Karras, U., Kreck, M., Neumann, W.D. and Ossa, E.

"Cutting and pasting of manifolds; SK-groups", Boston, Publish or Perish 1973.

Kosniowski, C.

[1] A note on *RO(G)* graded *G* bordism theory. *Oxford Quarterly J. of Math.* 26 (1975), 411-418.

[2] Generators of the unitary $\mathbb{Z}/p$ bordism ring. *Bull. Amer. Math. Soc.* 82 (1976), 344-346.

[3] Generators of the $\mathbb{Z}/p$ bordism ring. *Math. Z.* 149 (1976) 121-130.

[4] Characteristic numbers of $\mathbb{Z}/p$ manifolds. *J. Lond. Math. Soc.* 14 (1976), 283-295.

[5] $\mathbb{Z}/p$ manifolds with low dimensional fixed point set. "Transformation Groups", (L.M.S. Lecture Note Series 26). Cambridge, Cambridge University Press 1977, 92-120.

Lang, S.

"Differential Manifolds", Reading, Mass. and London, Addison Wesley 1972.

Lashof, R. and Rothenberg, H.

G smoothing theory. "Symposia in Pure Math" Amer. Math. Soc. Vol. 32, 1977.

Milnor, J.W.

[1] "Morse theory" (Annals of Mathematics Studies No. 51.) Princeton, New Jersey. Princeton University Press 1963.

[2] On the Stiefel Whitney numbers of complex and of spin manifolds. *Topology* 3 (1965) 223-230.

Milnor, J.W. and Stasheff, J.D.

"Characteristic classes" (Annals of Mathematics Studies No. 76.) Princeton, New Jersey. Princeton University Press 1974.

Ossa, E.

Einige resultate aus der aquivarianten bordismustheorie. Preprint.

Palais, R.S.

[1] "The classification of G spaces" (Memoirs of the American Mathematical Society No. 36.) Providence, Rhode Island. American Mathematical Society 1960.

[2] Slices and equivariant imbeddings. "Seminar of Transformation Groups" (Annals of Mathematics Studies No. 46.) Princeton, New Jersey. Princeton University Press 1960.

Pulikowski, W.

[1] $RO(G)$ graded G bordism theory. *Bull. Acad. Polonaise des Sci.*, 21 (1973), 991-995.

[2] Coefficients of $\mathbb{Z}/2$ bordism theory indexed by representations. *Bull. Acad. Polonaise des Sci.*, 21 (1973) 997-999.

Spanier, E.H.

"Algebraic Topology", New York, McGraw-Hill 1966.

Stong, R.E.

[1] "Notes on Cobordism Theory" (Mathematical Notes) Princeton, New Jersey. Princeton University Press 1968.

[2] "Unoriented Bordism and actions of finite groups" (Memoirs of the American Mathematical Society No. 103.) Providence, Rhode Island. American Mathematical Society 1970.

Stong, R.E.

[3] Bordism and Involutions. *Ann. Math.* 90 (1969) 47-74.

[4] Equivariant Bordism and $(\mathbb{Z}/2)^k$ actions. *Duke Math. J.* 37 (1970) 779-785.

[5] All in the family. Preprint.

Wasserman, A.G.

Equivariant differential topology. *Topology* 8 (1969) 127-150.

Index

Attaching a G handle   44

Balanced Product   3
Bordism   7
Bordism exact sequence   11
Bordism ring, equivariant   5
Bundle bordism   15
Burnside ring   221

Chern classes   81
Critical orbits   33
Critical points   33
Cutting and pasting   47
Cutting and pasting G vector bundles   50

Divisible   74
$\mathcal{D}(M)$ 59
Dold manifold   6
Double of a manifold   59

Elementary abelian 2 group   22, 77
Equivariant bordism   5
Equivariant cutting and pasting   48
Equivariant handle decomposition   44
Equivariant Morse lemma   34

Equivariant SK groups   49

Euler characteristic   55

Euler's function   21

Exact sequence of bordism groups   11

Family of G slice types   8

Family of subgroups   10

(F, F') bordism   9

$\Gamma(H;U)$ 17

G bordism   5

G bundle bordism   15

G manifold   3

G module   3

Gradient like vector field   43

Grothendieck group   48

G slice type   8

G_x slice   8

Handle decomposition   44

Hessian   33

$H_{m,n}$-Milnor manifold   6

Injective module   74

Invariant Morse function   40

Irreducible G module   19

Isotropy subgroup   3

Milnor manifold 6

Morse function 40

Non-degenerate critical orbit 34

$O(H;U)$ 17

Orbit 3

Order on $St(G)$ 29

Orientable fibration 84

Pasting manifolds 47

$P(m,n)$ -Dold manifolds 6

Principal $\Gamma(H;U)$ bundles 17

Projective space bundle 24

$RP(R \times E)$ 24

$RP(R \times (E_0, E_1)$ 136

Serre spectral sequence 84

$S(H)$ - G slice types with isotropy subgroup H 152

Singular bordism theory 7

Singular n manifold 7

SK equivalence 47

SK invariant 199

Slice 8

Slice theorem 7

Slice type 8

Sneiden und Kleben 47

Stiefel Whitney classes 74

$St(G)$ - G slice types 28

Surgery 31

Surgery of type $[H, V_1, V_2; F]$ 32

Type 8

Trace of a surgery 32

Tuple convention 2

Unoriented bordism ring

Vector bundle of type $[H;$

$\mathbb{Z}/2$ - SK invariants 205

$\mathbb{Z}/4$ - SK invariants 214